RODALE PLANS

Solar Growing Frame

Grows Fresh Vegetables Year Round, Using Only Solar Energy

edited by
Ray Wolf

Rodale Plans
Rodale Press, Emmaus, PA

6   8   10   9   7   5

Editor
Ray Wolf

Writers
Jack Ruttle
Ray Wolf

Technical Illustrator
Frank Rohrbach

Carpenter
Gerald Hock

Horticulture Research and Development
Dianne Matthews
Robert Flower

Design, Production and Illustration
Rodale Press Marketing Art Department

Cover Photograph by Carl Doney

Rodale Plans
Rodale Press
33 E. Minor St.
Emmaus, PA 18049

Printed in the United States of America
on recycled paper, containing a high percentage of de-inked fiber.

Library of Congress Cataloging in Publication Data

Wolf, Ray

Rodale's solar growing frame.

1. Solar growing frames — Design and construction.
2. Salad greens. I. Rodale Press. II. Title.
SB352.7.W64   690'.8'9   79-25815
ISBN 0-87857-305-4

Table of Contents

Introduction

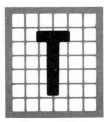his book is your passport to the world of sun power. Rodale's Solar Growing Frame uses the most modern solar heating techniques to produce fresh vegetables during the coldest months, without using any fossil fuels. A solar growing frame captures the sun's energy, and then uses the latest insulating and heat storage techniques to conserve every precious degree of that free energy.

The end result is a unit that literally coasts through the winter, keeping temperatures in the growing range for many vegetables. Thus, growth is slow, but sure, and a carefully planted and maintained frame will provide fresh greens every week of the year. In some extremely cold regions, growth may be suspended for a few weeks. But as soon as the weather begins to warm, the plants will continue their growth and you may continue your harvesting. For most areas, harvesting will continue without a break.

This book is your first step into a safe, sane and secure solar-based lifestyle, through the production of fresh food.

Actually, this is two books. The first part of this book serves as an owner's manual to the solar growing frame. It tells you how the frame works and how to use it successfully. The second part of the book will enable you to build your very own solar growing frame. The explicit step by step instructions, with illustrations and photos, are directly linked to the blueprints for clarity. The overall approach is that of a model maker. First, we'll help you make all the pieces to your solar growing frame, and then, we'll show you how you put them together.

We at Rodale are very excited about presenting plans for this solar growing frame. We know it will help provide people with fresh, good food, and encourage them toward greater self-sufficiency.

This project is the first of many projects aimed at bringing fairly sophisticated devices like the solar growing frame to the public, ensuring them of success, regardless of their building skills. The plans in this book have been designed and tested so that even with marginal carpentry skills you will be successful if the step by step instructions are carefully followed.

This book is our attempt at bringing solar technology to a reality for thousands of people. In the past, the grow frame idea would have remained only a good idea, but not a reality. With these plans, you now have access to a solar growing frame, using the blueprints and the instructions in this book you can take an active role in solarizing our lifestyles.

Solar growing frames have been a very important part of the research effort at Rodale Press and the Organic Gardening and Farming Research Center for the past two years. Over 20,000 temperature readings have been made at our research site, all recorded by hand. In all, ten different designs have been tested. By carefully using the data, we have included elements we know will work in Rodale's Solar Growing Frame.

Primary horticulture researchers have included Diane Matthews, Eileen Weinsteiger and Steve Ganser. Robert Flower has spent countless hours with the thermal data and Dr. Dave MacKinnon was instrumental in the early start of the project. Jack Ruttle has carefully worked with the project since its inception, and industrial designer Jim Eldon designed the first solar growing frame. This book reflects all of their hard work. Without their constant research in some of the worst winters our area has ever known, there would be no solar growing frame.

The actual preparation of the plans for the growing frame is mainly the work of Frank Rohrbach, Gerald Hock, and myself. Any errors in this book fall on my shoulders, not those who helped in its preparation.

We are confident here at Rodale that solar growing frames are the wave of the future. So much of your daily food budget is directly linked to the energy cost of moving perishable produce across the country from warmer areas to colder areas. The growing frame makes that use of energy, and outlay of capital unnecessary. Now you can produce your own fresh and delicious vegetables during the winter months, thanks to the sun—and Rodale's Solar Growing Frame.

Ray Wolf
Editor

THE SOLAR GROWING FRAME

The spark that ignited the solar growing frame fire began four years ago when, to test a newly emerging concept, the Organic Gardening and Farming Research Center built one of the country's first passive solar greenhouses. That initial greenhouse performed so well that even its supporters were somewhat surprised, and the skeptics were downright amazed.

Without the use of any external energy, the greenhouse maintained temperatures well within the growth requirements for a good number of crops, especially salad crops. To gain an appreciation of how well this greenhouse works, consider a week during February 1978. The average minimum air temperature outside the greenhouse that week was 3° F. Inside, the average low air temperature was 45° F. On the coldest night, temperatures outside plunged to a severe –17° F. That same night the greenhouse hit a low of only 38° F. In fact, the greenhouse averaged daily high temperatures in the 60s. But, even more important, soil temperatures at the root zone never dipped lower than 48° F.

The design of a solar greenhouse is simple. First, only the areas with glass are directly exposed to the sun. The entire north wall and north part of the roof are opaque, as are the major parts of the east and west walls. All walls have at least six inches of fiberglass insulation, and are tightly caulked to prevent infiltration of cold winter winds. The north wall is covered with five gallon cans painted black and filled with water. The sun shines on these cans, warming the water during the day. At night, the warm cans give off their heat to keep the rest of the greenhouse warm. All glass surfaces are double glazed and have an insulating panel to install at night, to reduce heat loss.

The thermal performance of the solar greenhouse was so encouraging that we began to look around to see how we could improve on the overall concept. Soon we were looking at the cold frames used at the experimental farm, asking how we could make those frames perform as well as the greenhouse.

At the same time our curiosity got aroused, a solar architect in New Hampshire, Leandre Poisson, unveiled his design for an insulated cold frame. Similar to the greenhouse design, this cold frame presented solid insulated walls to the north, east and west, and a solid glass face looking south, straight at the sun.

Because Poisson's design promised to do a lot of the things we wanted to test, we built one of his units. At the same time, Dave MacKinnon, the designer of our greenhouse, began to look at the new idea of solarized cold frames.

The major concept we wanted to try with our new generation of cold frames was the use of thermal mass, one of the keys to the greenhouse's success. Thermal mass is simply a large amount of material used to store heat. The two materials most often used for heat storage are masonry or water. Thus, we wanted a cold frame that incorporated a large amount of mass that would be warmed during the day, and would radiate that heat back to the plants

The Stone Pac used slabs of concrete for heat storage.

at night, protecting them from freezing.

We knew we would have to insulate the thermal mass from the outside elements, and we finally decided on the design shown below. The entire north wall was made up of thermal mass, insulated, with earth bermed around it to further moderate winter temperatures. In

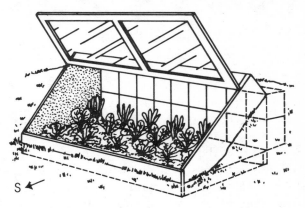

The Water Pac, a bermed coldframe with cans of water behind the growing area.

addition, we put a small amount of insulation around the growing bed and along the sides, with a glass door facing south. We built one frame with masonry for heat storage, and another using water. In addition, we built the initial unit designed by Poisson. We also built a very simple, A-frame cold frame with a solid plywood north wall and a glass south wall. The only insulation this frame had was one layer, ten inches deep around the perimeter of the growing bed.

The Can features movable insulation to protect against night heat loss.

That first year we had mixed results. Some of the things we found were extremely encouraging, others were surprising. Some crops we were sure would perform well did not. Others did better than we thought. The effect of constant cold, moisture, and the hot sun, took more of a toll on the construction of the frames than we had anticipated.

In all, we learned many lessons but the most important was the simple fact that al-

though there were problems, at times these prototype cold frames performed almost as well as the greenhouse. We knew we were onto something.

The following year, we had a total of ten frames ready for testing. We made some modifications to some of the first year's frames and improved our temperature monitoring techniques. But our biggest job was finding a way to get added heat into the frames, and to keep the soil warmer. Knowing that around the

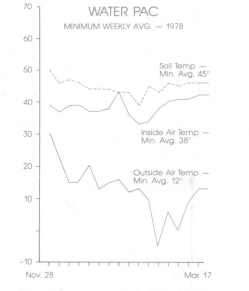

Thermal performance of the Water Pac.

turn of the century, French market gardeners used the heat of decomposing manure as a heat source for a cold frame we decided to experiment with the concept.

From our work the first year, we learned that the soil temperature inside the frame has

more effect on the air temperature inside the frame, than do outside temperatures. We decided to extend the depth of the insulation around our new frames in hopes of retaining more heat in the soil. Soil can absorb a lot of energy, making it a good source of thermal mass. But to be effective, thermal mass should be insulated from the elements.

One winter, thanks to an early snow covering, the frostline penetrated only about a foot deep at the research site. We knew we

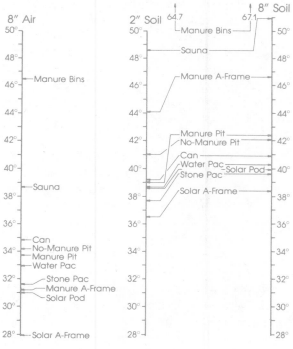

Thermal performance of grow frames.

wanted to have soil temperatures inside the frames no colder than the low 40s, or the plants would not grow well. We began to dig deeper and record the soil temperatures. Not until we probed three feet deep did we find soil at 39° F.

That finding told us that to achieve the necessary warmth, we either had to dig more than three feet deep, or insulate. We decided to compromise, digging a soil cavity two feet deep, and insulating it not only on the sides, but the bottom as well. Doing this makes a

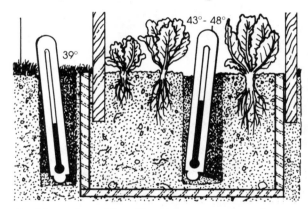

Inside and outside soil temperatures depend on the depth of insulation.

sealed box of soil. It benefits from the temperature-moderating effects of the surrounding soil, yet is insulated from the heat-robbing properties of the cooler soil. The insulation at the pit's bottom is installed in pieces, allowing for drainage.

Using this system, we achieved an average soil temperature of 43° F. near the surface, and 45° F. at eight-inches deep. These temperatures are sufficient to grow several crops. The

lesson we've learned is that cold earth extends about two and a half feet deeper than the frostline. With that factor in mind, you must either extend the insulated side walls to that depth, or use insulation across the bottom of the soil bed at a shallower depth.

If our Research Center were in an area with a deeper frost line, we would increase the amount of insulation from two inches to three inches. Likewise, in areas without a permanent frostline, we could reduce the insulation to one inch, and would not need the bottom insulation.

Going into our second year with the frames, we also began to use several oriental varieties of vegetables. Most of the types we were using were from the cabbage family. However, they don't taste like cabbage, they resemble spinach and other leafy-green salad crops. The most attractive feature of the oriental crops was their temperature requirements. They seemed almost perfectly designed for the temperatures we were getting in the frames our first year.

By the time winter set in the second year, we had ten cold frames in operation and set for testing, in addition to the solar greenhouse. All the frames tested had approximately a four-foot by eight-foot growing area. The opaque walls on all but one frame were insulated with two inches of styrofoam. All frames were double-glazed. Eight used an outer glazing of fiberglass and an inner layer of polyethylene. Two others used double layers of fiberglass. All growing beds used two-inch foam perimeter insulation, but to varying depths. Here is a rundown of the ten frames we monitored.

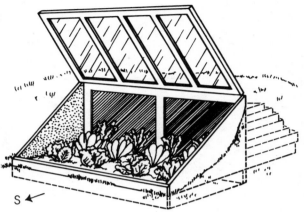

THE STONE PAC

The north wall of this frame is built out of three-foot thick concrete slabs, totalling 72 square feet of concrete. The wall is insulated and bermed on all sides except the south. The soil bed is insulated to a depth of 10 inches around the perimeter. This was one of the first units we tested, and was designed by Dave MacKinnon.

WATER PAC

Designed exactly like the stone pac, this unit substitutes water cans for concrete slabs as the method of heat storage. The north wall had a total of 180 gallons of water for heat storage.

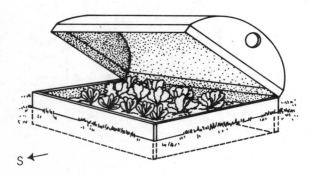

THE POD

Shaped like a large flat bubble, the pod is the only unit tested without an opaque north wall. It is made of two layers of curved fiberglass that rest on top of a seven-inch raised growing bed. The pod can be lifted completely off the frame when not needed. Insulation extends ten inches into the ground around the perimeter. The insulated ground is the only thermal mass in the frame. This unit was designed by Leandre Poisson.

THE SAUNA

Completely above ground, the sauna uses hot air solar collectors to augment the heat of the frame. The one-foot-deep growing bed is about three-and-a-half feet above ground level. Two convective hot air collectors, two-feet by two-feet, heat air during the day, warming 18-cubic feet of rocks below the soil bed. Collectors heat the rocks during the day. At night, the rocks heat the soil. The unit is covered with an insulating shutter to cut heat loss at night.

THE CAN

Leandre Poisson's original design, this unit uses insulated solid walls on three sides, with a two-layer fiberglass south wall. Its unique feature is a night insulation system, that is operated by turning a switch allowing styrofoam beads to fall between the two layers of glass. In the morning, the lid is raised and the beads run back into the storage can, where a turn of the lever keeps them for the day. We had problems when our unit got wet, and condensation formed between the two layers of glazing, and the beads did not flow. When the unit worked, it worked very well, maintaining good temperatures. We insulated ten inches into the ground around the unit, adding nine five gallon cans of water for more thermal mass.

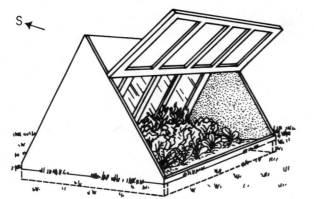
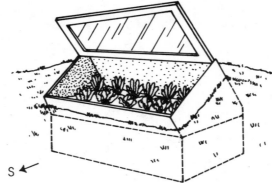
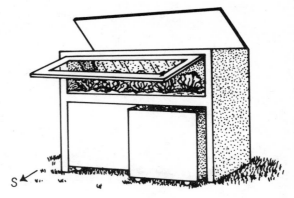

THE A-FRAME

The simplest of all designs we tested, this inexpensive unit uses uninsulated solid plywood north, east and west walls, with a clear south wall. The only thermal mass in this unit was the soil in the growing bed. The soil was insulated ten inches below ground around the perimeter.

MANURE A-FRAME

Located against the south wall of a building, this unit used a pit three-feet deep, insulated and filled with two-feet of manure and one-foot of soil, putting the growing level at ground level. For this frame, our main interest was to determine the benefits of putting a frame against a building.

MANURE PIT

A four-foot deep hole was dug and insulated on the sides and bottom. Two feet of manure were added to the hole and covered with one foot of soil, putting the growing level one foot below ground level. The top and sides were insulated and bermed.

NO MANURE PIT

This frame was built exactly like the manure pit, to serve as a control to gauge the effect of the manure. It held no manure, and was only insulated to a depth of two feet. The growing bed was a foot below ground level, and the soil bed used only one foot of soil.

MANURE BINS

With this frame, our goal was to find how to get heat from manure throughout an entire growing season. This unit is completely above ground and uses moveable bins of composting manure as a heat source. As the composting process of changing manure to compost is completed, the manure begins to cool. Then, the bins can be reloaded—without disturbing the plants—and the heating continues. This unit has a well insulated one-foot-deep growing bed. The entire unit can be shut tightly at night to cut down on heat loss.

DESIGN OF THE GROWING FRAME

From our experiences that year with ten experimental frames, we learned a lot that has had a major impact on the design of the growing frame you are about to build. Rodale's Solar Growing Frame is a lot different than any design we tested during the last two years. It takes the best of each, combining them into one unique frame.

As you'll see, this unit has outgrown the term *cold* frame. It is now truly a solar *growing* frame. Thanks to tight construction, solar principles and insulation, this frame is capable of promoting plant growth all during the year in most areas. The information that follows is a point by point rationale for the design. If you are thinking of making any changes, read this section carefully to understand fully why the unit is designed the way it is.

INFILTRATION

Infiltration of cold air is the quickest way to cancel out all the heat saving and collecting measures you may otherwise employ. Moving air carries heat away rapidly, quickly chilling the plants. The only way to eliminate infiltration is to make the frame tight. Pay careful attention to proper cutting measurements when building the frame, and give the entire unit a thorough caulking. Rodale's Solar Growing Frame is designed with a minimum number of exposed joints to reduce the chances of infiltration.

A traditional cold frame suffers from excessive heat loss.

The main thing we've learned in our two years of experimentation is that it's tougher to build a frame that will hold up to cold weather than we originally thought. For that reason, the new design calls for a cement block foundation, with no wood in direct contact with the ground. We've found that frost can knock a frame out of kilter, allowing winter winds easy access to plants. The best thing you can do to reduce infiltration is to put your growing frame on a foundation that will not move. If the foundation stays put, the frame will, too. The base must therefore be below the frost line. However, if you can't put in a below ground foundation, Section II shows a way to build the entire unit above ground.

WALL CONSTRUCTION

We've found that walls require at least two-inches of plastic foam insulation, equal to R-10. But, two-inches of foam is difficult to work with, especially when using standard size lumber. Thus, we've increased the width of insulation to 2½ inches, using 2 x 3 lumber for the walls. To get 2½ inches of foam, we combined a sheet of one-inch foam with a sheet of one-inch-and-a-half. Both sizes are usually available, or can be ordered by a lumber yard. If you cannot get the full-two-and-a-half inches of insulation, you can use two one-inch strips, leaving a half-inch air gap in the wall.

At our Research Center we tested different types of wall construction, finding that a standard stud wall, with ¼-inch plywood sheathing on both sides works best. The plywood protects the insulation from sunlight, errant shovels, and other garden tools. A tight plywood/foam sandwich with no air gap is stronger. But a wall with an air gap will be

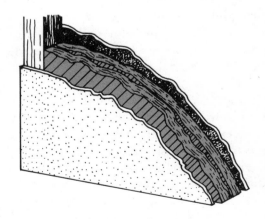

A wall sandwich has the insulation in direct contact with the inside and outside plywood skins.

strong enough if you do not berm earth against it.

There is still some discussion about whether the air gap should be towards the inside or outside of the frame. We recommend it go on the outside wall of the frame, because most moisture is inside the frame. The combination of paint, plywood with glue layers, and foam will effectively prevent the moisture from getting to the outside wall, where it may condense. If you have an air gap, watch the walls for any sign of swelling. Should they swell, put a small vent at the top and bottom of each stud cavity, venting only the air gap, not the entire frame.

INSULATION

While we've used styrofoam insulation throughout on our frames, many of the new, solid board urethane foams will also work. The entire frame does not have to be insulated with the same material.

For the foundation, either a urethane foam or styrofoam should be used. White bead-board types of insulation will not work when exposed to the weight of the soil, and the moisture. They simply come apart over time and lose their insulating value. Use a type of insulation that is approved for below ground use.

For the walls, we recommend you stay with either styrofoam or the urethane foam boards. Solid boards are easy to cut and install, and resist moisture well. The white bead-board type of insulation will work fine in the walls, but it does not have as high an R-value as the other types of board insulation. R-value is a means of indicating a materials resistance to heat loss. The higher the R-value, the better an insulation the material is. Fiberglass batts or cellulose types of insulation should not be used because of the potential moisture accumulation inside the walls.

The design of Rodale's Solar Growing Frame calls for an R-12.5 wall system (2½ inches of foam) and foundation insulation of R-7.5 (1½ inches of foam) beneath the growing bed and around the foundation. For other locations we recommend:

Milder Areas: for zone three of the climate map where freezes barely penetrate the ground you can probably omit the bottom insulation of the growing bed. Above ground wall insulation could be reduced to two-inches of foam, but the foundation should be insulated around its perimeter.

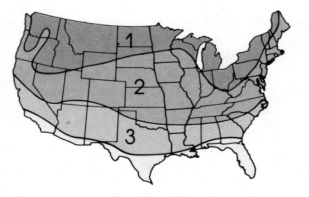

Temperate Areas: for zone two on the climate map where the frostline extends from about one-foot to four-feet deep, make no changes.

Cold Areas: for zone one on the climate map, where frost penetrates more than four feet, add one-inch of foam to all below ground insulation levels, and give thought to changing to 2 x 4 walls with a full three and a half inches of insulation.

INSULATING SHUTTER

The insulating panel that covers the south face at night is an invaluable part of keeping enough warmth inside the frame to support

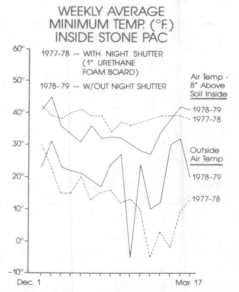

The benefit of a thermal shutter, compared to no shutter.

vegetable production. In the dead of winter, more than two-thirds of all heat loss occurs at night. And the double-glazed, clear south face loses about two and a half times more heat than all the rest of the frame. In our tests, insulating panels or blankets generally kept air temperatures ten degrees warmer and soil temperatures three degrees warmer than where no night insulation was used.

The nighttime insulation is just as important as insulation in the walls, or double glazing, or insulation below ground for heat storage in making a growing frame work. A good, simple shutter that seals in heat at night will keep the unit warmer than an extra inch of foam in the walls or below ground. An inch added to the walls increases their insulation value by only one-third, but a shutter increases the south face's insulation value by a factor of four or more.

A nighttime insulation system that is durable, draft-proof, water resistant and easy to operate is not easy to design. At night, our shutter system is folded down from the inside rear wall where it is stored during the day. Having the shutter inside reduces the weathering problem enormously. Furthermore it is easier to make airtight because it doesn't have to withstand wind pressure. Although there are plenty of alternatives you can try, bear in mind that the shutter we recommend will only cost about fifteen dollars, and should boost your temperatures from five to ten degrees. It's a good investment.

The material we use for the shutter is a urethane foam coated with aluminum foil on both sides. Once the material is cut to fit, all edges are taped with PVC tape to protect the foam from sunlight and abrasion. The material we use is one-inch thick, is rated at R-8, and is known as Thermax Sheathing, made by Celotex. You should use material at least one-inch thick, or the shutter will sag in the middle, and you may have to add a support brace.

One year we tried to use fiberglass insulation batts by wrapping them in polyethylene. We sealed the polyethylene with pop-rivets and ironed the seams with an old electric iron to melt the edges together. The curtain was waterproof and manageable for a while. And it did an excellent job of keeping the cold frames warmer. Its weight and flexibility kept drafts out quite well. But before the winter was two-thirds over the polyethylene degraded to the point of tearing readily.

European cold frame gardeners used to rely on mats made of reed or straw. Similar

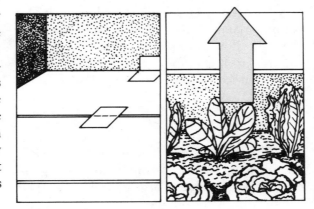

A thermal shutter holds in heat at night, plants without a shutter lose heat.

night insulation systems are still commonly used in China. If you can make something like this cheaply, do so. You will need long stalks of straw or marsh grasses. They are laced together at the top and bottom to make a mat. Dried leaves that are loosely packed in bags to make a pillow are another readily-available, low-cost material with a fairly high insulation value. People often bag leaves in poly for the trash. They're water tight but won't last through the winter. Poly-bagged leaves (loosely packed and dry) should be set into burlap sacks to protect them against sunlight and handling. Even though the burlap will become wet and frozen, the pillows should remain manageable most of the time.

During the cloudy periods—especially when it's very cold—the frame can be left with the shutter closed for several days with no harm to the plants. When you sense that the structure is losing heat and cooling down steadily because the sun can't recharge it, shutter the heat in and fend off a freeze. Without light the plants are in suspended animation at these cool temperatures. They'll be fine when you open the frame up to the sun again.

DOUBLE GLAZING

Glass is the traditional material used in cold frames. However, a growing frame is not a cold frame. Glass is difficult to cut until you learn how. Moreover, it is tricky to install, especially on angled surfaces like this frame, and can cost twice as much as some fiberglass used in greenhouses. If you are lucky, you can get glass free from old buildings or from aluminum storm window companies. On the plus side, glass will last forever, if not broken by accident. And it looks good.

Very clear plastics, like Plexiglas and Lexan are also good in cold frames, but generally quite expensive. There is a whole new range of double walled acrylic glazings designed for passive solar use. These materials are all good, but too expensive for a growing frame.

The glazing method we finally decided to use is comprised of an outer layer of high-quality fiberglass and an inner layer of polyethylene. Fiberglass is our choice because it is tough, long-lasting, easy to work with, and cheap. The fiberglass does not look clear but it is formulated to transmit the full spectrum of

Sunlight passes through glass and is absorbed inside the frame. The energy radiated by the frame does not pass through glass, creating heat build up.

light that plants need. Fiberglass has a tendency to yellow over the years, reducing its sunlight transmission somewhat, but the manufacturers sell a paint-on substance that prevents this degrading.

Polyethylene should not be used for the outer glazing. You lose the greenhouse heating effect with an outer layer of polyethylene. Heat-producing radiation (long wave) passes back out through polyethylene quite readily, but glass or the other solid plastic glazings reflect that long wave heat radiation back into the frame increasing the heat of the unit.

Even in mild climates the frame will require a second layer of glazing. One glazing layer has almost no insulating value. Glazing materials have poor R-values to start with and they are extremely thin. But two layers create a dead air space. The dead air space has an insulating value of approximately R-1.5. Plus, the inner glazing layer is another preventive against infiltration of cold air.

We recommend polyethylene for this second glazing layer. Used inside the unit, the poly should last several years before needing replacement and saves a lot of money compared to two layers of fiberglass.

SMALL DOORS

The door we designed in the south face of the growing frame is small, and for good reason. We found that large glazed lids warped severely over the course of a winter. Long spans of wood don't hold up well when the moisture levels inside and out are high and the

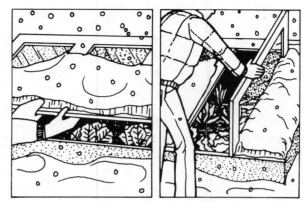

Small doors are much easier to use when winter snow and ice build up on the frame.

lids are being operated almost daily. Furthermore, the water that crept up and under the sill would freeze and force twisting. The result is cracks that let the wind blow in.

The shorter dimensions of the small door mean that any warping has a smaller effect. Moreover, small doors are easier to make more rigid and durable. And the potential crack area is greatly reduced. The smaller doors are easier on *you* in the bargain and they let far less frigid air in on the vegetables when you open up the frame to pick or water plants.

SHAPE

The high, vertical north wall of the growing frame may be a surprise to many people, who think a lower, more sleek design would work better. Actually, the vertical wall is a very important part of the design. This one feature allows the amount of clear glazing to be larger,

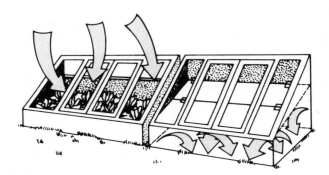

The vertical wall helps catch energy during the day. At night the shutter seals in the heat.

and at an angle where more direct radiation from the sun can be captured during the winter months.

Angled glazing captures about twice as much solar light as would a horizontal glazing of the same size. The glossy white inside surface of the walls reflects all the captured light around the frame and onto the plants. At night, with the shutter in place, the parts of the frame that have growing plants do have the long, low look.

We tested frames with sloping rear walls, and vertical walls. We found that the frames we tested captured from 40 to 60 percent of the light outside the units. The vertical-wall units were constantly in the 60 percent of available light category, while those with sloping walls were in the bottom rankings. Plus, angled walls are harder to build and support, especially if the unit is to be bermed.

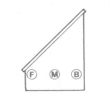

LIGHT INSIDE AS % OF LIGHT OUTSIDE

DOUBLE GLAZING 45° FRAME

	F	M	B
	Front	Middle	Back
Very clear sky	82%	78%	87%
Clear sky	83%	89%	97%

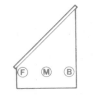

LIGHT INSIDE AS % OF LIGHT OUTSIDE

SINGLE GLAZING 40° FRAME

	F	M	B
	Front	Middle	Back
Very clear sky	100%	98%	116%
Clear sky	101%	110%	118%

Earth berming helps to moderate temperature extremes, but places structural stress on the frame.

BERMING

Berming means mounding earth up around the outsides sides. A berm isolates the cold frame walls from temperature drops that occur in the air most nights. The temperature around the walls never goes any lower than the earth in the berm. The berm maintains an even temperature compared to the air which can plunge 20° F. or more below the average for a freak night or two. Don't berm wooden-walled cold frames. We tried it on several cold frames and found the added weight bent wood walls. The wood will rot much more quickly as well.

Instead, try piling hay bales all around the frame, or locating the frame on the south face of a building. Either method can be like adding another three-quarters-of-an-inch of polystyrene foam, about R-4, to the walls of the frame.

HEAT STORAGE

The large volume of moist growing bed inside the frame's insulated shell doubles as heat storage or thermal mass. Each cubic foot of that soil absorbs or releases about 45 BTU's (a measurement of heat energy) for every one degree rise or fall in its temperature. In the winter months at our experimental farm in Maxatawny, Pennsylvania we get about 50 percent of the maximum possible sunshine. Nightly low temperatures in midwinter average between 10 and 20° F. Yet the two-inches of foam insulation and the energy collecting ability of the frame keeps the soil temperatures inside between 38 and 42° F. We have found that temperature range warm enough to keep cold-tolerant, leafy vegetables producing all winter.

If you can achieve similar minimum soil temperatures in your climate, then your frame won't need more heat storage. But you may want warmer temperatures to make vegetables grow faster. Soil temperatures between 50 and 60° F. will produce more rapid growth of the winter vegetables.

Besides too-cool soil temperatures, you can tell if you need more heat storage if air temperatures become too high during the day. For winter vegetables, the air temperature shouldn't go over 80° F. If the frame overheats, you will have to vent it to the outside, and that is wasted heat. Ideally, that extra energy should go into heat storage. We almost never have overheating problems in our frames during winter, but from February on, it is very

HEAT STORAGE CAPACITIES OF COMMON MATERIALS

(BTU's per cubic foot per one degree of temperature change)

Material	BTU
Water	62
Soil	43
Rock or Brick	25
Wood	23
Sand	18

possible. On a clear day, we have to stay alert for a heat wave in our growing frame. The tighter you build your frame, the more likely it will be to overheat.

By stacking more heat storing materials on the north wall, more of the energy inside the grow frame can be held for later use. Solar greenhouses, from which our frame evolved, depend on this principle for most of their heat. Light that hits storage materials on the north wall is absorbed by the thermal mass instead of being bounced back to contribute to overheating.

The best storage materials are rock (earth and other masonry materials are also good) and water. Water can hold more than three times the heat of an equal volume of rock or brick. But both kinds of material work very well and are cheap, if not free. In the growing frame the best system is one that does not take up much

of the valuable growing surface and which extends high up the north wall to get above plants' leaves and into full sun. Brick is perfect for a tall three-inch thick stack. For water, aluminum soda or beer cans (sealed with tape) are stackable and will occupy about the same volume as bricks. Secure either of these walls by stretching chicken wire across it so the stack won't fall. Another good stackable container is a one gallon rectangular paint-thinner can. Even round paint buckets will do. The containers should be painted a dark color; black is common but very dark red or blue work nearly as well.

We have used massive north walls of both water and stone in our growing frame experiments, and have been surprised to learn that neither has much of a thermal improvement over similar frames without them. We are not sure why because there were a great many variables in these experiments, as they had

Thermal mass gathers heat during the day, releasing it to the plants at night. A shutter holds heat in the growing area.

been set up. It could be that the frames with mass were losing heat to infiltration faster than the others and the storage materials made up for it. Another possibility is that much of the energy that the mass walls radiate at night is going straight out through the glazing (we operated without thermal shutters the year we ran these tests). We just don't know at this point. But the concept has been proven to work very well in the solar greenhouse and it should work in the growing frames, too. Locating your thermal mass so it is covered by the night insulation would improve the effectiveness of the mass as shown in the illustration.

HOT COMPOST OR MANURE

Our experiments with composting manure proved that it is a very effective source of supplemental heat, as tradition suggested it would be. We found that hot compost in a pit below the frame can raise soil temperatures at least into the low fifties—ten degrees higher than in frames with soil only.

For some people there will be drawbacks to using hot compost or manure. Large amounts are required and it may not be easy to get where you live. And we learned that the manure should not be put into a pit to compost, which means a special cold frame must be designed for efficient use of manure. Composting manure in a pit requires an excavation four-feet deep (rather than one just two-feet deep) in order to accommodate two-feet of compost. That is a lot of work, and it must be repeated once each year in order to charge up your growing frame. Moreover, we found that the manure in the pits finished giving off heat about seven weeks after the initial charge—in late January. There was no way to recharge it and after that the system worked no better than a soil-only frame.

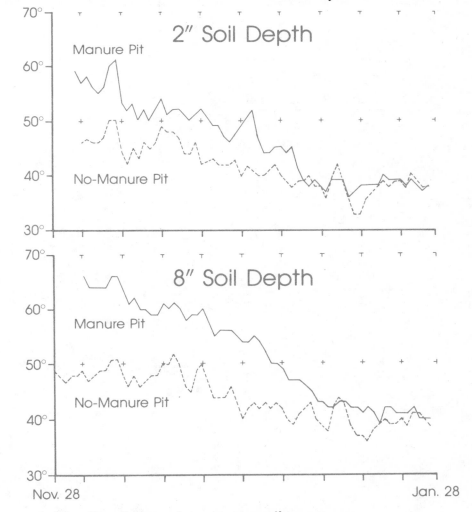

The effect of composting manure on soil temperatures.

PICKING A SITE FOR YOUR GROWING FRAME

 **M**ore than anything else, a growing frame needs sunshine. Satisfying the growing frame's solar needs is the most important consideration to picking a site, but there are other things to weigh.

During the normal outdoor gardening season it is very useful to have the frame close to the garden. On the other hand, if the garden is far from the kitchen door, consider a location nearer the house. During the winter months when you're depending on the growing frame most, you won't have much call to visit the garden.

Locating the growing frame on the south side of a building or a sheltering windbreak, hedge or wall will significantly diminish the heat sapping effect of winter wind. Growing frames, painted to match the house, look good nestled against a south-facing wall. Additionally, the windbreaking effect of your home will boost temperatures inside the frame a valuable degree or two at no expense.

The growing frame's site should be well drained. If at all possible, the two-foot-deep foundation of the frame should be completely underground for a permanent installation. But that two-foot foundation should be well above the water-table at all times of the year if it is to last. Plants will not grow well if their soil is water-logged. If frigid waters invade the frame's soil, the plant roots will be severely stressed and the soil temperature levels will be reduced below that needed for growth. Tem-

peratures once lost will be regained very slowly. Be sure the site you pick will not be subject to runoff water from thawing ice or melting snow. See the foundation instructions in Section II for information on above ground growing frames, for those with drainage problems.

Ideally, the solar growing frame should receive full sun from 9:00 a.m. to 3:00 p.m. during winter months. Achieving this requires two things. You need to determine the direction of true south. And you must be sure that no buildings, trees, or hills will cast shade on the frame's energy-collecting, clear face in winter. The sun is low in the sky at that time of year and obstructions can cast far-reaching shadows through the morning and afternoon. If you are siting your frame during a season other than winter, you'll have to calculate how far those shadows will reach.

When orienting your growing frame,

Push a stick into the ground, checking with a level to be sure it is straight.

remember a compass does not point to true south, and you want your frame facing true solar south. While there are special maps that show how much the magnetic south shown on a compass differs from true solar south at various spots all over the continent, there's a more direct, and easier way.

The sun reaches its high point above the horizon—the position of most intense energy reception on earth—at 12 noon every day of the year. That's one o'clock daylight savings time. Go out to the potential frame site a little before noon on a bright sunny day and shove a thin, very straight rod or stick firmly into the ground. Use a level to make it perpendicular to the ground in all directions. At noon standard time (1:00 p.m. daylight savings time), the shadow of the rod will point straight north-south. Mark the north end of the shadow with another stick and you are set. This line should be parallel to the side wall of your frame when it is installed.

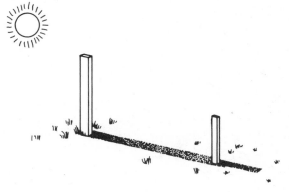

At noon, insert a stake at the top of the shadow.

Use a carpenter's square to mark a line perpendicular to the true south line and mark it with pegs and strings. That will be the east-west axis of the growing frame, and should be parallel to the front wall of the finished frame.

Deciding how far heat robbing winter shadows will reach is a little harder. Shadows stretch their longest on the days around December 21. You can watch the shadows at that time of year to see where to locate the frame but for other times of the year, sun path charts, while somewhat tedious, will give you approximate answers to the shadow question with just a few minutes of calculations.

Sun path charts contain a lot of information. On one calculated for your location, you can plot the sun's position for any hour of the day, any month of the year. Imagine that on the sample sun chart, point A is your growing frame site. Of the seven arching lines, the one closest to point A is the sun's path in December. The line closest to point B is the sun's path in

June. The other lines are for pairs of months, paired: May/July; April/August; March/September; February/October; and January/November.

The vertical dotted lines are the hours of the day. Noon, point B is in the center, sunrise point C is to the left, and sunset point D is to the right.

Now that you have an idea of how to read

a sample sun chart, look at the eight sun charts shown and find the one for your locale. Any road map will tell you at what latitude you live. Find the sun path chart for your latitude and get out a pencil.

Using the blank sun path chart provided on page 27, you are now ready to plot the sun path for your location for the month of December. First read the angle where the De-

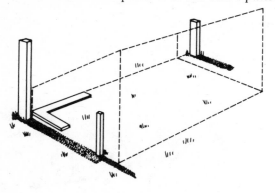

Use a carpenter's square to make a perpendicular line from the first line established. This will mark the front line of the frame.

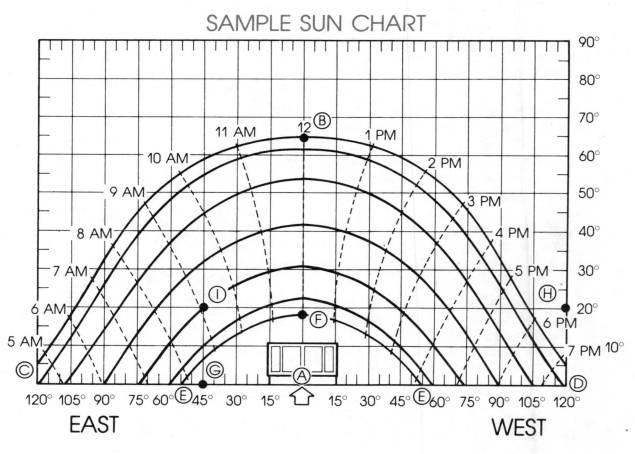

SAMPLE SUN CHART

cember sun path goes off the edge of the chart for your location. Both sunrise and sunset should register the same angle. Find it and mark it on the blank chart along the horizontal bottom line. Looking back at the sample sun chart, these two points are point E. Next, read the angle for 12 noon on the chart for your location, and mark it on the vertical scale, in the center of the paper. This is point F on the sample sun chart.

That may sound confusing, but all you end up with is three marks. Two run across the bottom line of the graph, the other on the vertical noon line.

Draw an arch connecting all three points, and you have the sun's path over your growing frame location for the month of December. Right now you do not need to plot any other monthly paths.

You are now ready to go out to the potential frame site, and mark any potential shading obstructions on the graph paper. To

record where a potential shade producing object is, you need to know two things. How far in degrees it is from true south, and how many degrees on the horizon it rises. To do this, you need a straight edge, protractor and a piece of string with a washer tied on the end of it.

To mark how many degrees off true south an object is put the 0° axis of the protractor along the north-south string you marked the frame location with. Next, place the edge of a straight edge on the line, and move the south end of the straight edge in line with the obstruction. Read the angle it differs from true south, and record that angle on the blank chart's horizontal axis. Continue to do that for all potential obstructions.

Next, you have to calculate how high on the horizon these obstructions rise. Tie the plumb line into the center hole of the protractor, and sight along the protractor's straight edge to the top of the obstruction. The plumb

line will cross a number on the outer edge of the protractor giving you the approximate altitude of the obstruction. Plot this along the vertical axis of the chart. Be sure you hold the protractor at eye level when doing this sighting work, or the readings will be off a few degrees.

Plot the two marks as one, and mark it with a dot. For the obstruction shown in the illustration, the plotting on the sample sun chart would look like points G, H combined as point I. If the plotted mark does not touch the line representing the sun's path, then the object will not cause shading problems. If the mark does touch a line, it will cast a shadow on the growing frame for the months represented by the line, and all other months below the line.

If it looks like an object may just cause a little amount of shade, go back to the sun path chart for your latitude, and plot in the January/ November sun path and see how much shadow you get for those months.

Lightly sketch in hour lines onto the blank chart as shown on the chart for your location. This will let you know during what hours of the day the shadow will be a problem. You are mainly only concerned with the hours when the sun is the strongest.

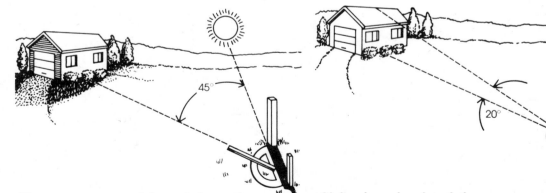

Use a protractor to sight any obstructions, noting their angle from true south.

Sight along the edge of the protractor, reading the altitude of the obstructions where the string crosses the protractor scale.

LATITUDINAL SUN CHARTS

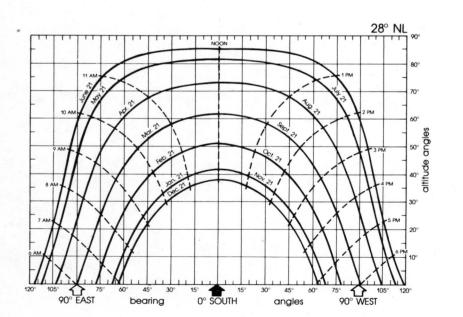

28° NL

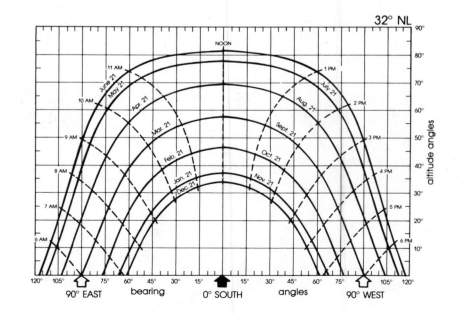

32° NL

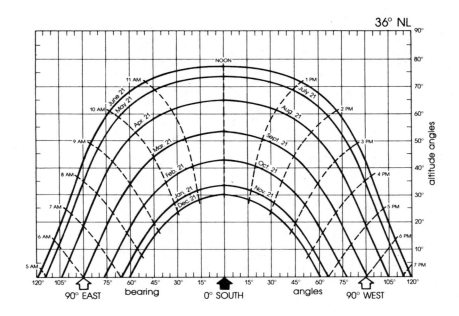

36° NL

altitude angles

90° EAST   bearing   0° SOUTH   angles   90° WEST

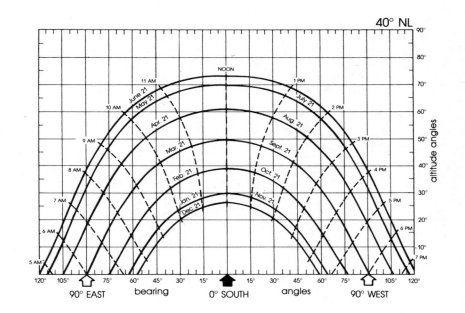

40° NL

altitude angles

90° EAST   bearing   0° SOUTH   angles   90° WEST

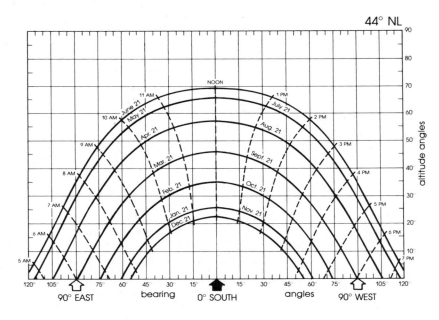

44° NL

altitude angles

90° EAST   bearing   0° SOUTH   angles   90° WEST

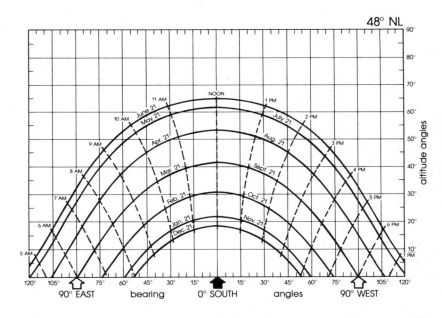

48° NL

altitude angles

90° EAST   bearing   0° SOUTH   angles   90° WEST

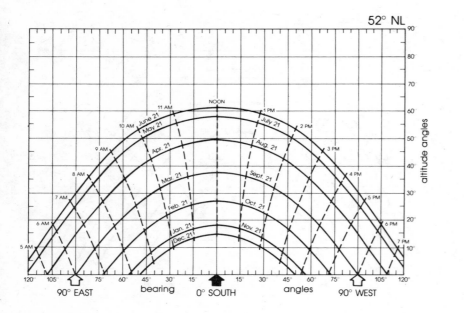

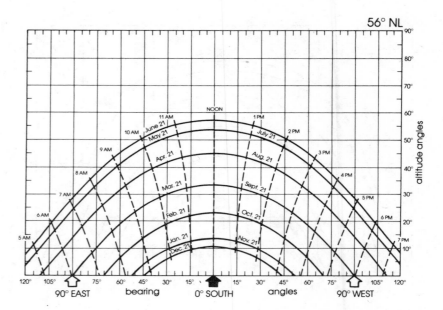

Check a road map to find the latitude for your area. Then consult the sun chart for that latitude to find the plotting points to fill in the sun chart for your site. You need not plot every monthly chart, only the December and June charts will be needed.

These sun charts are reproduced from *The Passive Solar Energy Book* by Edward Mazria, with the permission of the publisher, Rodale Press, Inc., 33 East Minor Street, Emmaus, PA 18049.

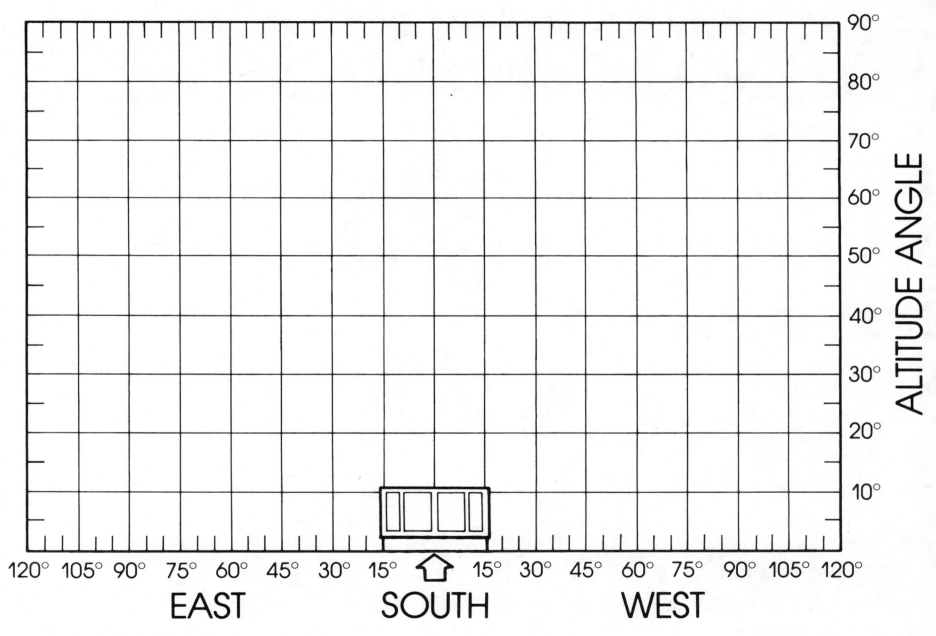

ALTITUDE ANGLE

90°
80°
70°
60°
50°
40°
30°
20°
10°

120° 105° 90° 75° 60° 45° 30° 15° 15° 30° 45° 60° 75° 90° 105° 120°

EAST   SOUTH   WEST

GARDENING IN A GROWING FRAME

hen you think of gardening in a growing frame, the thing you have to remember is that the climate inside the frame won't be like a summer day during the winter months. What you will be doing is moderating the winter climate until it is more like spring or fall. If you've ever seen a good spring or fall garden, you'll have some idea of how beautiful and productive a well-handled growing frame can really be.

To give you an idea of just what is possible, let's look at the yields from our ten growing frames and the passive greenhouse last year at the Research Center. Remember, these plantings were experimental and were not as dense as could be, and some non-cold-hardy varieties were included. The average yield from the group of solar devices was 14.7 ounces of greens per square foot of growing surface during the four month period of January through April. The growing frame you'll be building has just over 27 square feet of surface.

If you do as well as our average, you'll have the potential of producing just under 25 pounds of fresh greens during the winter. An average large salad takes about two ounces of greens, giving eight salads from one pound of greens. That gives your one growing frame the potential to produce as many as 200 fresh salads during those four months.

Keep in mind that those figures are an average of all frames we tested. Some did far better than others. If you do the same calculation based on the top performing growing frame, you get a potential yield of 40 pounds of vegetables, or 320 salads from one four foot by eight foot growing frame.

The frame you will be building has better glazing, a tighter door fit, and a much better foundation, plus increased insulation in the walls. Potentially, it should thermally outperform any of the frames we tested except those with added heat from manure or solar collectors. Due to the frame's improved design features, if you do add manure, you should be able to get better thermal performance than anything we've yet reached at our farm. Our current experiments with added heat and extra thermal mass in frames like yours promise great results.

One last word of caution about potential yields. Remember that the figures given are a four month average. January is the toughest month for growing frame gardening. The temperatures are almost as cold as February, yet the light levels are less. As the days begin to get longer, and the frame receives more sun-

Vegetables should be continuously harvested, concentrating on the outside leaves.

shine, production will really take off. If you have your frame growing at its peak in November and December, you'll come into January with well-established crops, and they will hold well. Unless you have warmer weather than we have, you won't get much plant growth during January, but you will still be able to harvest regularly. Solar growing frames may be able to do wonders, but they can't produce miracles.

Lastly, the yield calculations are mainly with the specialized salad crops discussed later on. You can expect similar thermal performance from the growing frame you'll build, it's just a question of how long it takes you to learn how to intensively garden the frame.

When you read our vegetable recommendations, you'll see that a lot of the crops do best when transplanted into the frame. The ideal setup is to have the solar growing frame producing food outside, and a small seedling area inside the house, getting plants ready to transplant into the frame to keep production levels up.

During our first experiences with growing frames we were so intent on getting through the winter that we failed to get the frame stocked up for spring. As a result we found the units almost empty by April with ideal growing conditions inside, while in the outside garden we were still waiting to plant. Keep a constant succession of plants going into the frame, and it will keep a steady stream of salad greens on the table twelve months a year.

For the fall season in a growing frame, start with seedlings grown indoors or sheltered from summer's hot and dry conditions that restrict germination and healthy growth. In midwinter start with house-grown transplants because the frames are too cold for efficient sprouting. Most salad vegetables germinate best in soil that's 65 to 70° F. The soil should never be dry, always moist. These conditions are most easily provided indoors for much of the year. Plus space is too valuable in a growing frame for nurturing seedlings. You want to use every inch of space for the production of food.

Some of the vegetables you'll be growing, especially lettuce and the brassicas, seem to do better if they are transplanted as soon as they fill their container, which may mean two times before they are big enough to go into the growing frame. Transplant them into small pots when they show their first true leaves. Transplant them again when they have about four leaves and are about three inches tall (this will differ greatly between varieties).

Feed the seedlings about once a week with a very weak solution of fish emulsion or some other fertilizer. The goal is to raise replacement transplants that will be large and vigorous enough to compete with the larger plants already established in the growing frame. For the first planting at summer's end, you can set much smaller plants directly into the frames. But from then on, stick to good-sized transplants.

The most important thing to remember about gardening in a growing frame is that the solar energy available to plants in April is the same as in August. You know how well things grow in August, in the garden, but why not in April? The answer is in the warmth of the soil. In April the soil is just thawing from the winter, and takes time and energy to warm up—outside. In a growing frame, your April soil is as warm as the garden's June soil. Toward the end of the winter season, start replacing the cold season crops with efficient users of space that like warmer weather. Then in the fall, reverse the process, and begin to replace warm season crops with cool season crops. For those extremely cold areas, you can use the frame during your warm season to grow crops, like cantaloupes, that you've never grown before.

One thing you'll love about gardening in a growing frame is the lack of insect problems. We have not had pest problems in our test frames, most likely because of the cool environment. In fall and spring, aphids pose the largest potential problem. To avoid any outbreaks, be careful to only transplant clean stock, checking the underside of leaves for bugs before moving transplants into the frame. A few hitchhiking aphids will quickly start a population boom that will give you headaches for months.

If during spring you notice the beginning of a soil-borne insect problem, let the frame's heat-gathering potential go to work for you. During the summer when the frame is either empty or being used as a food dryer, shut the unit tight, and let the sun superheat it for a couple of days. Stir up the top six- or eight-inches of soil, and shut it for another period of baking. The high heat will kill off most of the insect eggs and larvae with no trouble.

The frame is such an efficient heat gatherer, that overheating during the growing season is sometimes a problem. The ideal air temperature for cold hardy vegetables (and many other plants too) is in the low 70s. In fall and late winter, the air temperature in the frame can push into the 80s and higher on clear days. Overheated air should be vented if the incoming air is not so cold that it will chill the plants and if the soil bed has accumulated enough energy to promote good growth. Once the daily high soil temperature (at six-inches deep) reaches the mid-50s, you can begin venting hot air out of the frame. When the ground is colder, the frame really can't afford to sacrifice the energy that is lost with the hot air. The latch on the frame is designed to hold the door open, while preventing wind from blowing it wide open.

During times of extremely cloudy and cold weather, you may want to leave the frame covered with the night shutter day and night until sunny weather returns to recharge the frame. On very cloudy days there may be so little sunlight getting through that the frame loses heat all day long. If the weather appears likely to persist, consider leaving the frame closed up. The plants go into suspended animation and can be held that way for several weeks. On any day the sun is shining, the frame will charge up, even in the coldest weather.

A deep freeze is a highly unlikely event if your growing frame is well built. But if the plants inside the frame do freeze, don't try to artificially thaw them. Most cold-hardy vegetables don't freeze until the temperature drops well below 32° F. And they can withstand those temperatures for quite some time. After the leaves thaw out, they are fine and the plant keeps on growing. But repeated freezing and thawing is damaging. Not much growth can occur anyway if the temperature is low enough to freeze these plants. So if the plants freeze once or twice they will recover. And if everything in the frame should freeze up solid for a stretch, leave it alone until the weather turns nicer.

If the ground inside the frame freezes, the plants will be unable to get water and their leaves must be shielded from sunlight. After the weather warms, irrigate the frame with warm water. Then cover the glazing with lath or some other material that will reduce some of the light. This will prevent extreme overheating while the soil and plants thaw.

It's all right to pick vegetables on cold days. Very little energy is contained in the warm air that is lost, and heat stored in the soil will quickly re-charge the fresh, cold air to above freezing temperatures. But do try to work quickly and get the frame locked up tight as soon as possible. Ideally, pick during the morning and solar energy will heat the air back up before nightfall. Try to avoid picking on very cold days.

Always harvest mature outer leaves and pick just one or two leaves gradually from all the plants. Pick leaves when they begin to touch and shade neighboring plants. Always leave a little stub of leaf stalk on the plant.

This way you don't risk tearing tissue off the stalk or leaving any other pathway open to decay organisms.

This will keep each plant at maximum health and vigor. If you pick all the mature leaves in the frame, give the plants time to regrow rather than start to eat the young leaves. If you notice the stem beginning to elongate and the leaf flavor begin to turn stronger, harvest the entire plant. It has reached its peak and you should replace it with a new young plant. So that you never waste space or time in the frame, always have young transplants ready to go in as soon as you remove an old plant.

SOIL FOR THE GROWING FRAME

The soil in the bed of the growing frame merits more attention than any other patch of garden. The soil needs to hold moisture without becoming water-logged. It should be rich in nutrients to promote the plant growth. And it should be rich in organic matter both for healthier plants and for the generation of carbon dioxide. When filling your foundation, you should replace at least the top 12 inches of soil with a special blend. Ideally replace all two-feet with the blend.

The soil you make for the growing frame should be an equal mix of well-made, mature compost and good topsoil. If the native topsoil is clay, lighten it with sand. If the soil is sandy

and drys out quickly, make the final mix a little richer in compost.

Compost is the perfect material for growing frame gardening. Compost, of course, contains nutrients by itself and is an especially valuable source of trace minerals. Plus soil rich in compost also helps control certain bacteria and fungi that prey on plants because it promotes the growth of a wide array of beneficial microorganisms that attack the damaging microbes. Compost also has the ability to hold on to plant nutrients and maintain them in a form that plants can use. So since you'll be feeding the plants fairly often, compost helps make the use of nutrient solutions much more efficient.

During the winter, you will need to occasionally water the soil in your growing frame. If the soil stays wet after watering, root rot may occur. Use compost or sand to help drain soil. Compost is best because spaces between the particles contain air that waterlogged soils lack. With compost, plant roots can breathe, even though there is moisture in the compost around them. Additionally, the spongy particles of organic matter contain enough moisture to prevent the growing frame from drying out too fast.

Carbon dioxide, the gas plants use to make nutrients through photosynthesis, can be used up very quickly on a sunny day in an airtight structure like the growing frame. But as soil microorganisms digest the organic matter in the warm inch or two of soil near the surface, they generate more carbon dioxide. Compost is the only economical source of adequate carbon dioxide for a growing frame.

Compost naturally has a pH of about 6.5, exactly right for growing salad vegetables. Both lettuce and the brassicas require plenty of calcium. If your soil needs more, add a bit of dolomitic limestone, but watch that the pH doesn't rise over 7. Again, compost will help buffer the effects of an addition of limestone, keeping the pH near 6.5.

To make good compost for the growing frame, you'll need two kinds of ingredients: nitrogen-rich materials like manure, grass clippings, or seed meals of cotton, sunflower, soy and the like; and carbon (or energy-rich) materials like hay, straw, or dried leaves. Mix an equal amount (by volume) of carbon-rich material with an equal amount of nitrogen-rich material. Then give it time and a few stirrings and you'll have good compost. Rapid and complete decomposition depends on the materials having enough nitrogen and air. If you turn it once a week (for air), the compost should reach 140 to 160° F. (pasteurizing temperature). If it doesn't, add more nitrogen. Turning weekly will provide finished compost in a month, provided the mix is properly balanced. The pile should be spongy moist, never wet. Start with twice the volume you'll need for your frame soil bed, because compost decreases in volume by about half as it digests.

As long as the soil in your growing frame is cool, (under 50° F.), you will have to provide the plants with extra nutrients. When the soil temperature inside the growing frame slips below 50° F., the microbes that release nutrients from the compost to the plant roots are nearly inactive. By using nutrient solutions ("manure teas") you can compensate for the natural loss of fertility in winter by providing the elements the plants need in a form they can readily absorb. However, remember that plants absorb much less nutrients and water at cool temperatures.

Fish emulsion and liquified seaweed are commercial products that are well suited to midwinter feedings. Fish emulsion is especially good because it contains a lot of calcium in addition to nitrogen, phosphorus, and potash in fairly good proportions to each other. Liquid seaweed is most valuable as a source of trace minerals. It is a little low in phosphorus. Don't use it as a primary source of nutrition because it tends to be high in salt, which will accumulate in your soil.

Blood meal, wood ashes, seed meals (soy, cotton, sunflower), fresh manure and dried manure are all good raw materials for making your own liquid concentrate. Their nutrients dissolve fairly readily and are in a form the plants can readily use. Blood meal is very rich in nitrogen but contains no phosphorus or potash. Wood ashes have no nitrogen but are a fairly good source of potassium and an excellent source of other minerals. Seed meals are pretty well balanced, but a little low in phosphorus. About half the nitrogen and phosphorus in manures will dissolve in water. Generally, manure teas are high in nitrogen and low in phosphorus.

When using nutrient teas, be stingy. Make very diluted solutions. Use them often rather than giving your plants one or two massive doses. For the soil and the growing conditions at the Organic Gardening and Farming Research Center, we've found that once a week seems to be enough. To compensate for the difficulty in getting phosphorus in a nutrient solution, sprinkle some bonemeal evenly on the soil around the plants and let the watering soak it into the soil. When soil temperatures rise back into the 50s, cut back on the rate of feeding. The best guide is to keep an attentive eye on the plant growth and yield. Too much nitrogen, combined with cold soil will cause plants to wilt, as if they need water. If your plants wilt, yet the soil is moist, skip the next few feedings.

You will not need to water much in midwinter, generally no more than once a week. At times, especially during cloudy weather, the plants will require much less water. Always water first thing in the morning on a day that promises to be sunny. Use water that's 60 to 70 degrees. If you notice salts from your nutrient solution building up on the surface of the soil, you should flush them away with a heavy drenching (but don't do this in the coldest months). This is one reason that you need extremely well-drained soil.

CHOOSING THE RIGHT VARIETIES FOR YOUR GROWING FRAME

We recommend you stick to growing cold hardy salad vegetables exclusively in your frame during cold months for several reasons. Most important, they grow in winter conditions. Many kinds of vegetables won't grow at all when temperatures hover below 50° F. most of the time. And that will be the case inside a growing frame in any location where no outdoor gardening is possible in winter. In warmer climates, solar growing gardening is much easier and a wider variety of plants than we discuss here can be grown with no loss in productivity. In fact, in some southern locations a growing frame may be too warm for cold hardy crops to do well.

The yield from salad vegetables is high because you eat almost everything the plant produces. Plants that set fruit or produce edible roots or stems require a much larger energy input from the plant compared to salad crops. As a general rule, stay away from any plants in this category if you want maximum productivity from your frame. Tomatoes and other heat loving crops are not the only poor choices for growing frame gardening. Many cold hardy plants make poor use of space and should be avoided. Peas must make a lot of vine growth before the first edible pods become ripe. The same is true of carrots, beans, Brussels sprouts, and broccoli. They all thrive when it's cool and can survive temperatures well below freezing. But winter light levels are too low for them to produce their edible parts efficiently. In a growing frame, they will be spindly, and fruits or edible roots will be small.

Salad greens have very little energy content for us as food. Salad plants, therefore, can produce leaves under relatively low energy conditions. Salad plants are extremely valuable as foods, however, because they are high in minerals and vitamins. And very often the minerals are in just the right form and proportions that our bodies need. All the vegetables in the cabbage family are unusually rich in sources of calcium. Many cold hardy plants are much higher in vitamin C than other vegetables. All the dark green vegetables are excellent sources of vitamin A. Vitamin A precursor (carotene) is destroyed rather quickly by cooking or just sitting around after harvest, so eating these plants fresh and raw provides some of the best food you can get in winter.

Physically, salad plants are ideal for a growing frame. They are shallow rooted, making the frame's two-foot deep insulated soil bed more than ample for vigorous root growth. They don't grow too tall, so they don't shade other crops as much as taller varieties would. Moreover, limited height keeps the plants below the level of the insulating night shutter. The front of the frame has about ten-inches of growing space at night, while the back of the frame has about 18-inches of growing space, more than enough for salad crops. Keep these measurements in mind when planting, as well as shading considerations.

As for temperature, salad crops are the perfect choice for a growing frame during the cold season. Temperatures maintained in the frame give most salad plants optimum growing conditions. Leaves grow very succulent, thin and tender in the cool temperatures and low light levels of midwinter, making them even better eating. Vegetables we sometimes think of as coarse and bitter, like kale, chard and endive, become sweet and tender when forced into midwinter production in a growing frame. Many of the crops that do best in a growing frame are ideal in the kitchen when used for stir-fry, steamed in soups and other ways besides salads.

There's plenty of variety too. We've grown more than two dozen species of plants. Best of them all, we think, are the many varieties of Chinese cabbages. These are not coarse like most European brassicas. There are many smooth leaved varieties with long succulent leaf stalks. They are very well suited to midwinter growing frame conditions. After all, Chinese plant breeders have been selecting vegetables that will thrive in the large unheated pit greenhouses that Chinese market gardeners have been using for centuries. Almost certainly, they will be new to you. But you'll be pleasantly surprised as we were by their lettuce-like qualities and sweet, delicious flavor.

Here are the vegetables we like best of all those we've tried.

PAK CHOI (or BOI CHOI), the White Cabbage.

(Brassica Chinensis, var. Chinensis).

A very juicy and sweet vegetable with crisp leaves and stems. Stem-pieces in salad have the appearance and texture of celery, with a superior flavor. The plant begins producing edible leaves rapidly, can withstand temperatures as low as 22° F., but will bolt very early in spring.

It can grow to form a loose head ten to 14 inches tall. The inner stalks are white, while the leaves are green, somewhat like Swiss chard. But for steady salad production it is better to continually harvest outer leaves. Under this method of harvest, the plants should be spaced closely, about six inches apart.

PAK CHOI (CHOY SUM), the Flowering White Cabbage.

(Brassica parachinensis).

We recommend this variety because of its flavor, rather than its productivity. It bolts to flower under winter conditions. But it's supposed to. The entire plant is harvested just prior to flowering. The tender, sweet leaves and stalk are chopped for salads or stir-frying. It consequently requires frequent replantings through the winter. The plants can be spaced quite close together because they grow up, not out. A dense planting can overcome its low per-plant yield.

SEPPAKU TAINA.

Another loose-heading Chinese cabbage, it is both extremely tasty and productive. The leaves are big, very smooth and dark green. The wide oval leaves grow at the ends of rather narrow, pure white stalks. Whether in salad or stir-fried, the stalks are as good as the leaves. The bunch of leaves grows fairly upright, and if left unpicked, the plant will form a two-pound, celery-like bunch of leaves and stalks about 14-inches tall.

SEPPAKU TAINA will survive air temperatures in the low 20s and will make good growth at 45° F. The best way to use it is to pick off outer leaves as soon as they expand to full size and allow new ones to form.

Set two- or three-inch seedlings into the ground about five- to six-inches apart, and keep them picked back to avoid crowding.

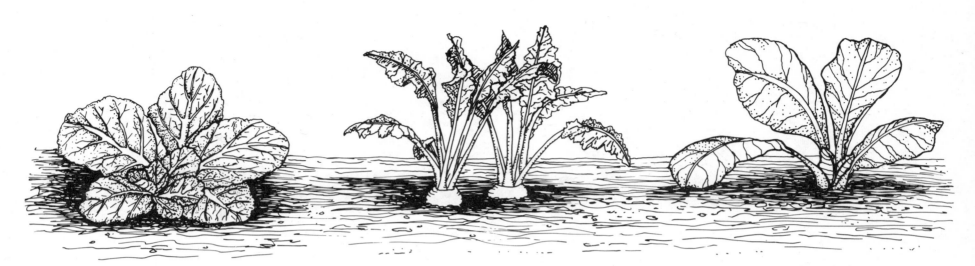

SIEW CHOY,
Yellow Bud Radish.

(Brassica pekinensis var. cylindrica).

Again, this is a loose-heading Chinese cabbage which will form an open rosette of leaves. Its leaves are light green, faintly savoyed and hairless. Its flavor is very mild, but with a distinct hint of mustard. Plant six-inches apart and keep harvesting outer leaves or the plant will eventually form a ten- to twelve-inch head.

TURNIPS, TOKYO CROSS,
PETITE WHITE, EXPRESS WHITE.

These and radishes are the only root crops that grow fast enough in a cold frame to be worthwhile. We like these varieties better than radishes because the globes are so sweet and they grow better. They are crisp and sweet, almost never bitter. Harvest them when they reach two or three times radish size. The skin is so tender that they never need peeling. The leaves are hairless and can be used in salads or stir-fried. Plant seedlings two-inches apart or seed them directly into the frame in fall and thin them as you harvest.

KOMATSUNA.

Another Chinese brassica, it has dark green, oval leaves on long white stalks. It is one of the most tender and mildest tasting. Although it is one of the slowest growing varieties we've tried, it is also very cold tolerant. Space the plants about eight inches apart.

Chinese Style Cooked Greens

1 tablespoon vegetable oil
1–2 cloves garlic, minced
1 onion, sliced
1 green pepper, chopped
4 cups chinese greens, coarsely sliced
1 cup mung bean sprouts
1 tablespoon tamari soy sauce
1 tablespoon water

Heat oil in heavy skillet or wok. Add garlic, onion and green pepper and stir fry a few minutes. Stir in oriental greens and sprouts. Mix soy sauce and water and add to vegetables. Cover and cook over medium heat until vegetables are just tender, about three to five minutes.

Yield: 4 servings

LETTUCE,
ARCTIC KING.

SPINACH,
MONNOPA

KALE,
DWARF CURLED

Some varieties of lettuce do very well under cold conditions and others do very poorly. Arctic King, a loose-head type, is the best we've tried. Its leaves are light green and very crinkly. Other good varieties are Green Ice, Ostinata, Kagran, Salad Bowl, Butter-crunch and Continuity.

Lettuce seed germinates best between 60 and 68° F. With fresh lettuce seed, light also helps sprouting. Transplant seedlings when they get four leaves, about four weeks after seeding. Do not set the crown of the young plant below ground level or it may rot. Make sure the pH is not lower than 6.5.

Because spinach is one of the earliest, most cold tolerant garden crops in the spring, we expected spinach to be one of the best winter vegetables in a growing frame. Most of the spinach plantings we tried, however, have been disappointing. The plants just don't grow well until spring. Monnopa is the only variety we've had success with every season. We plant spinach in flats and transplant the seedlings one to a peat pot while they are very small. We harvest outer leaves as soon as they seem to have reached full size.

Kale is the only european brassica that is well-suited to winter production in a growing frame. We were surprised by its excellent flavor. It grows very sweet and tender, not at all like the kale you pick outside in fall. And it grows very rapidly in early and late winter giving high yields all spring.

Dwarf kale is a good variety because it doesn't crowd other plants. It will be relatively tall, however, and should be planted next to the north wall of the frame where it has the most vertical space. Kale seed should be started in August or September to have plants of good size by the time harvesting starts in late fall. If you bring garden-grown kale plants into the frame, be sure to clean off all aphids.

CHICORY, MAGDEBURGH SUGAR HAT. Italian chicory.

(Cicorium intybus).

The leaves look like very large, thin, and delicate dandelion leaves. The plant sends out new leaves quite vigorously in winter, but after a while it is prone to mildew. It can be a bitter salad green, similar to endive. Start seeds in August, and transplant six-inches apart in the frame.

ENDIVE.

Endive is fairly productive in a growing frame. As the plants get old it becomes even more bitter, but grow frame grown plants are less bitter than garden grown plants. Seed should be started in late summer. Set plants six-inches apart.

PARSLEY.

Any variety grows well in the cold frame. The Italian broad-leaved variety is sweeter and more suitable for salads, but the curled varieties are more compact, and do better in a grow frame. The seed should be planted very early. Pot up a few tiny plants from your normal garden seeding and raise them in ever-larger-pots until you are ready to put them in the frames. Parsley plants grow slowly but one or two plants should suffice.

KYO MIZUNA.

SWISS CHARD.

**GAI CHOY,
India Mustard.**

(Brassica juncea, var. foliosa).

This is the strangest Brassica you'll ever see. Its leaves are very feathery and grow in a rosette. You could most easily mistake it for endive. It grows very fast, and while it is not as heavy a producer as other varieties, it tastes very good and makes a salad *look* interesting. Space the plants at least eight-inches apart.

Chard does best when it's brought into the growing frame as an already established transplant. Either start it in the frame early in summer or transplant mature plants into the frame a month before frost. The stored energy which a mature plant has assimilated helps it keep producing leaves vigorously well into the winter. Be sure that you are not bringing any aphids or other insects into the frame with the plants.

This vegetable has a more pronounced mustard flavor than any other of the oriental brassicas we've tried. It forms a very loose head and grows very fast. Grow it like the others, with six-inches between plants.

Cold Korean Green Salad

1 pound fresh greens	Steam greens in small amount of boiling water in a covered saucepan. When barely wilted drain in a colander and gently squeeze out excess liquid. Finely chop greens. Combine remaining ingredients and toss greens with dressing. Chill until serving time.
1 tablespoon salad oil	
2 tablespoons toasted sesame seeds	
1 clove garlic, minced	
1 tablespoon tamari soy sauce	
1½ teaspoons vinegar	
	Yield: 4 servings
	contributed by Joanne Penny

MICHIHILI.

Seed for this Chinese cabbage is one of the most widely available varieties. It grows fast and produces a lot of food. The leaves are more hairy than most people like in salads, but like the other Chinese cabbages, it's great in oriental dishes. Space plants six-inches apart.

SHUNGIKU, the Garland Chrysanthemum, or Chop Suey Greens.

The leaves, of course, look just like small chrysanthemum leaves. One- to two-inch long shoots are tender. The flavor is unusual and mildly pungent. It produces steadily, though slowly, all winter.

Many more traditional salad crops that we've tried do not grow well in the growing frame. Beet and carrot seeds were able to germinate, but the plants never grew beyond seedling stage. Sugar peas, fava beans and kohlrabi, grew slowly and became healthy but rather small plants. They never reached a harvestable stage all winter.

We have quite a few vegetables and varieties yet to try. The Chinese cabbage family seems to be the biggest potential source of really cold-hardy vegetables. There are lettuces that have been bred specifically for winter production in unheated greenhouses in Holland and England. We suspect that corn salad will grow well in growing frames. Chervil and chives are both cold-tolerant in the outdoor garden and are delicious in salads.

Good sources of Oriental and European cold hardy vegetable varieties are:

Herbst Bros.
100 North Main St.
Brewster, NY 10509

Kitasawa
356 W. Taylor
San Jose, CA 95112

Tsang and Ma International
P.O. Box 294
Belmont, CA 94020

Thompson & Morgan
Box 100
Farmingdale, NJ 07727

Steamed Chinese Greens

1 pound oriental greens
2 tablespoons oil
1 clove garlic, minced
1 onion, chopped
2 tablespoons vinegar
1 tablespoon molasses

Steam greens in a small amount of boiling water in a covered pot. Drain thoroughly. Heat oil in small skillet and add garlic, onion and saute until golden. Add vinegar and molasses and combine thoroughly. Add greens, mix well and heat through.

Yield: 4 servings

Snappy Tossed Salad

4 cups chopped greens particularly spinach, kale or chard types
½ cup minced onion
⅓ cup mayonnaise
2 hard-boiled eggs, sliced

Stir the greens and onions together. Toss with mayonnaise and garnish with eggs.

Yield: 4 servings

WATCHING YOUR GROW FRAME

The last thing you have to do to be a successful grow frame gardener, is watch and record how your frame performs. All the subtle differences you gloss over in the garden are magnified in a growing frame. We recommend you record every day's temperatures for the frame, as well as the outside air. To do this best, a min/max thermometer for inside the frame is ideal. Be sure to place the thermometer below the night shutter to get a reading of the plant's climate, not the colder outside. An outside recorder of min/max temperatures will also give you a good idea of how well your unit works, and how extreme the weather can get before you have to be concerned. Lastly, a soil thermometer will enable you to take occasional readings to see how the soil is doing.

On the chart provided, record the readings you take. There is a column for calculating the difference between the minimum inside the frame, and that outside the frame. This will give you a good indication of how well the frame is working. By noting what type of day it is, you will soon begin to learn how overcast a day can be and the frame still heats up. Likewise, you'll learn what type of days to leave the shutter closed, and not subject the frame to cold temperatures. The only way you can learn to really garden the frame successfully is to learn how it performs thermally. To learn the frame's thermal performance, you must keep some type of temperature log.

We have only supplied one chart, to record the frame for one month. Have some photocopies made of the page, and record temperatures for at least one year, if not more. After that time you should begin to know the frame and how it will perform.

You also should pay careful attention to what you plant when, and how it does. Most likely you will be trying new varieties and you will have to decide if you like their taste, and if you want to grow them again. Keep careful notes until you become completely familiar with this new generation of plants. If you have any insect problems, make note of them and what you think may have caused the problem.

As with the temperature data, record your gardening information on the form provided, or make up your own form. If you want to use the one provided, have photocopies made and keep the completed forms in the back of this book. That way you'll always be able to look back and see how your frame performed.

You'll find that your growing frame actually develops somewhat of a personality. Soon it will begin to be like a family pet, and you'll go out at night to pull the shutter in place, "to keep it warm and cozy." In the morning, you'll go out to lift the shutter, "just to say goodmorning." If you think it helps to talk to plants, wait until you see what talking to a growing frame does for both you and the frame.

It is important to carefully watch and monitor your frame. Not only will the frame be supplying you and your family with food, but it will be a learning experience. This most likely will be your first formal foray into the world of solar energy. Monitor how well your frame does, and soon you'll be looking at ways to apply solar energy to other heating needs around the house. A solar growing frame is your first step on the journey to tomorrow. Have a good trip.

Month		1	2	3	4	5	6	7	8	9	10	11	12	13	14	15	16	17	18	19	20	21	22	23	24	25	26	27	28	29	30	31
Soil Temp.																																
Max. Temp.	Inside																															
	Outside																															
	Difference																															
Min. Temp.	Inside																															
	Outside																															
	Difference																															
Description: (Cloudy, Windy, etc.)																																

Variety				
Seed Source				
Date Seed Started				
Planting Method				
Date Transplanted				
Date Harvest Started				
Date Harvest Ended				
Performance				

Now that you know what a growing frame is, and what it can do for you, it's time to learn how to build your very own solar growing frame. The following step by step instructions are designed to enable even those with very limited carpentry ability to successfully build a growing frame. For some, the instructions may prove to be overly simple, but for others, it should tax your abilities to the very limit.

The key to success is to follow the instructions to the letter. Any changes you may want to make to one piece will have ramifications for other parts of the frame. If you want to change some part of the design, carefully study all the blueprints, until you fully understand how interrelated all parts of the unit are, and calculate what effect on other pieces your change may have.

Unless you are a very good carpenter, we suggest you forgo any changes. The design is well thought out and tested. The plans themselves have been tested. We know if a complete novice follows the directions, a successful solar growing frame will result. We've investigated most of the changes you most likely will think of, and they've been rejected for one reason or another. Trust us, the design and the instructions are good.

Before you begin, an understanding of how the plans are designed to be used will help you through the project. The entire operation is designed like making a model. The first thing you do is cut out all the pieces for the frame, and then proceed to put the pieces together. All ten pages of the blueprints are lettered. Every piece on every page is given a number, that combined with the letter of the page gives every piece a label. Thus, if you see a part numbered G-3, go to page G of the prints, The Foundation, and see what part G-3 is, and you find that it is an anchor bolt to hold the sill (G-5&6) to the concrete blocks (G-1), and that it is fastened in place with mortar (G-2). Thus, at a glance you can find any piece, what size it should be and how it should be cut or used.

The flow of work is designed to have you doing other things, while paint or preservative is drying. Thus, the first step is to cut all the wood for the entire project. Follow the cutting section carefully, and you will get a complete growing frame out of the lumber listed. All sizes have been designed for economy of materials, and the cutting diagram is the most efficient way to cut them. As you cut each piece, mark what letter and number it is, and put it aside. After cutting all the wood, we recommend that you treat the wood with a wood preservative that is not toxic to plants. Many commercial preservatives are toxic, so follow the advice in that section carefully.

After the wood is treated, and drying, we recommend you put in the foundation for your growing frame. We explain a new type of concrete block construction that does not need a foundation, and you don't have to put mortar between the blocks. For people who are easily intimidated by masonry, that may sound too good to be true. If you follow the instructions, you'll end up with a foundation that should outlast you.

For those with masonry skills, a standard block and mortar foundation will work just fine. In cases where a permanent foundation is not possible, you can build a freestanding unit. The changes you will need to make in the plans to build a freestanding frame are explained at the end of the foundation section. A freestanding unit has some drawbacks, mainly the question of how well it will hold up in years to come, but in some cases it will prove to be the only way to have a growing frame.

While the foundation is curing, we recommend you assemble the top frame and the doors. This is most likely the most tricky part of the project, and should be painted before the outer glazing is finally put in place. Thus, in between coats of paint, you should be putting together the four wall units of the frame. Then, go back and put the finishing touches on the doors and top frame. Lastly, put the four walls and the top into place on the foundation, weatherseal it and apply a good coat of paint, and you're ready for the growing season, Rodale's Solar Growing Frame style.

One of the problems with a book of this sort, is that the materials that are available to us in eastern Pennsylvania, may not be available to you. If you have a thorough understanding of what you'll be buying before you go to the store, you'll be better able to find suitable replacement materials. You'll have to use your own judgement and find substitutes when needed. However, we've tried to stay away from any special types of products, and when we do use one, we give the name, address, and phone of the manufacturer. If you can't find the material, call the manufacturer to find a local source of supply.

Much of the cost of a Solar Growing Frame is in the insulation. Unfortunately, you can not cut corners at this point. The growing frame works primarily because it can capture and hold heat. Without insulation, the frame would not be able to hold heat as well. Unfortunately, there are very few areas where you can cut costs in the frame. If you use salvaged lumber, be sure that it is the same as modern day lumber sizes. The chart on Sheet J will give the actual measurements of the lumber sizes we used in the plans. Be sure your salvaged lumber meets these measurements, or the frame will not go together correctly.

Whenever we felt a carpenters' trick may help you, we've passed it along. If a cut is potentially dangerous, we've noted it. Likewise, if the chance for error is high for a particular cut or operation, we've noted it and tried to explain it a second time.

This book should be right by your side in the shop. Every step is referenced to the blueprint. Make sure you know what you are about to do before you do it, and you should have little trouble. With the exception of one cut, everything on the growing frame can be done with portable power tools. Hand tools will also work, but would take longer. We've noted what tools you need, and what tools would be nice to have.

In our shop, we've found that it takes longer to treat and paint the 89 pieces of wood in the unit than it does to make the actual growing frame. With a competent carpenter, the project, start to finish takes no more than 20 hours, not counting the foundation. For a beginner, working weekends, figure to take from three to four weekends. That would break down as one weekend to cut and treat the wood. One weekend to put the top and doors together, another for the foundation and walls, and a last weekend to put the unit all together and paint it. The unit is built in pieces, and assembled on the foundation. This lets you work in the convenience of your basement or garage, wherever your tools are.

Be sure you have a helper when putting the frame together on the foundation. The walls and top are heavy, and a strong gust of ill-timed wind could easily damage the pieces. Having a helper is the only way to accurately put the frame together, especially fastening the top to the walls.

Lastly, we've designed a frame of permanence. In some cases there may well be a cheaper way to build the frame, but we've put a premium on durability, and suggest you do the same. A solar growing frame is literally a small house exposed to the elements all the time, with no expenditure of energy to help moderate the climate. It must resist infiltration, be weathertight, and offer plants the proper growing conditions. It must cope with extreme temperature shifts, and resist almost constant moisture and condensation.

Your solar growing frame calls for an investment of both your time and money. Build it well and it will last you many years, repaying your initial investment many times over. The climate created inside the frame is ideal for plants, but is hard on building materials. Watch for any signs of deterioration inside the frame, and be sure to give the frame a new coat of paint on a regular basis.

In short, a solar growing frame is no easy project to design. But with our plans, we're confident you'll find it not only easy to build, but it will work, and it will last. That means it will continue to provide your family with out-of-season greens for many years to come. Not a bad deal for a few weekends' construction time.

Construction Steps

CUTTING — all pieces are cut to length, the joints cut, and all wood treated with a preservative.

FOUNDATION — the hole is dug, leveled with gravel and cinder blocks put in place, without mortar. A surface bonding cement is applied and the foundation cures for a week.

TOP FRAME AND DOORS — these two units are put together at the same time, painted and then glazing put in place. Lastly, they are weatherstripped.

WALL ASSEMBLY — the four walls are put together, and the insulation cut to fit. The outside skins of the walls are not put on at this time.

ASSEMBLY — the four walls are put in place on the foundation, the flashing put in place, and the outside skins put on the walls. The top/door unit is then put in place, and the entire unit is painted and caulked. Lastly, plants and seeds are planted.

MATERIALS

The first step in building a growing frame, after reading this book, is buying your materials. Buying the lumber for the project should be easy, it's finding all the hardware that gets tough. First off, we recommend you take a hint from the phone company and let your fingers do the walking. Call around to local lumber yards to get price quotes and check on availability of the lumber. Very few places will give you price quotes for hardware over the phone.

To help you around the problem of unavailability, the use of each material is explained, and what characteristics it should have are explained. Whenever we've used a special product, we give you the name, address and phone number of the manufacturer. If you can't find the material in your area, write or call for the name of your regional distributor. Then contact the distributor and find what stores in your area they deliver to, and order the material from that store. Lots of times the store will not know they can order the material, so be sure to give the store the name of the distributor, as well as the name of the product you want.

LUMBER

2 x 8 — eight feet — four pieces
2 x 4 — eight feet — five pieces
2 x 4 — ten feet — two pieces
2 x 3 — eight feet — seventeen pieces
1 x 6 — ten feet — one piece
1 x 3 — eight feet — twelve pieces
¼ x ⅞ lattice strips — eight feet —
    six pieces
¼ inch exterior grade plywood, good one
    side — four feet by eight feet —
    four pieces

We recommend you purchase stud quality lumber. When buying 2 x 3 lumber, quality can often be questionable. However, with kiln dried studs, you normally will get much better wood. Buy all your wood from the same yard, as the thickness of the wood should all be the same. You may want to consider paying the premium price and hand selecting your wood. You will find construction much easier with straight, knot free lumber, especially with the 2 x 4's you'll need.

MASONRY

Concrete block — six-inch — 54 pieces
Surface bonding cement — 50 lb bags —
    two bags
Ready mix concrete — 50 lb bags —
    two bags
Gravel — half-inch size or smaller —
    one-and one-half cubic yards

These materials are for a foundation made with surface bonding cement. If you wish to make a different style foundation, read the text under Foundation. For information on the availability of Surewall Surface Bonding Cement, contact The INCA Co., Stanton & Empire Sts., Wilkes-Barre, Pa., 18702. 717:822-2191.

INSULATION

Foam — one-inch by two-feet by
    eight-feet — three pieces
Foam — one and one-half-inch by
    two-feet by eight-feet — thirteen pieces
Thermal Sheathing — one-inch by
    four-feet by eight-feet — one piece

As discussed earlier, other types of foam insulation will work, as long as they are rated for the way you will use them. For above ground insulation, you will need a total of three pieces of one-inch and three pieces of one-and one-half-inch insulation. If you buy two-inch insulation, the above ground frame only takes three pieces. Read the text for any changes to the thickness of insulation, and information about venting the wall. We used blue, extruded styrofoam from Dow Chemical, Construction Materials Division, Midland, Michigan,

48640. 517:636-1000. The insulation for the night shutter is Thermax Sheathing, made by The Celotex Corporation, 1500 North Dale Maybry Highway, Tampa, Florida, 33622. 813:871-4811.

GLAZING

Fiberglass — four-feet by ten-feet, .025 thick — one sheet

Polyethylene — 10 x 25 feet, six mil thick — one roll

We have used Kalwall Sun-Lite reinforced fiberglass for all our experiments. The thinnest they sell is .025 of an inch. This is fine for the growing frame. Kalwall .025 is sold in sheets 4 x 10 feet. We have designed the frame to be made from a sheet this size, if you carefully follow the cutting diagram. Several other companies sell reinforced fiberglass sheeting for greenhouse usage, and all will work just fine. Contact Kalwall Corporation, P.O. Box 237, Manchester, N.H., 03105. 603:668-8186. The other major manufacturer is Filon Division, Vistron Corporation, 12333 Van Ness Ave., Hawthorne, Ca., 90250. 213:757-5141.

A smaller size roll of polyethylene will work, but in this thickness, this is the standard minimum roll. Be sure to get clear, not natural color.

HARDWARE

Screws — 120 flat head brass screws — 1¼ inch #8

Stainless steel screws will also work, but are harder to find. Standard coated screws will rust after a few years exposure to the elements. The actual cost increase of brass compared to coated screws is not that much more, and not having to rebuild your frame in three years should make the difference worthwhile.

Nails — 12 d common nails — one pound
3 d finishing brads — ¼ pound
1¼ inch galvanized nails — two pounds

Try to get a barbed nail that has a good sized head. If you can't find galvanized nails of this length, you can go up to 2¼ inches in length. If you just can't get galvanized nails, get aluminum nails. Don't use standard nails. In the constantly moist climate inside the frame, nails will rust, staining the paint and losing strength.

Hinges — three sets brass 3 inch flat hinges
four sets brass 1½ inch flat hinges

Three sets of three inch open brass flat hinges to hinge the doors to the frame. Don't use the small screws that come with the hinges, use the 1¼ inch #8 brass screws already listed.

Four pairs of 1½ x 1½ inch open brass hinges to keep the two pieces of shutter together, and to hinge the shutter to the rear wall. The screws that come with the hinges will work fine, as long as they are brass.

Anchor bolts — ten bolts, ⅜ x 8 inches, with nuts and washers

Hanger bolts — twelve ⅜ x 4 inch hanger bolts

Hanger bolts have a woodscrew thread on one end, and a machine tread on the other end. We've looked all over and can not find these in stainless or brass. Thus, we use zinc coated. However, with a brass cap nut they should last quite a while, as they will be completely protected.

Brass cap nuts — twelve ⅜-16 brass cap nuts

Brass cap nuts — Two ¼-20 brass cap nuts

Brass hex nuts — ten ¼-20 hex nuts and washers

Threaded brass rod — two lengths, ¼-20 brass threaded rod, 12 inches long

Brass handles — four brass handles

Two fairly large handles to open the doors, and two smaller handles for the shutter. Most likely you will need to use larger screws than provided for the handles on the doors. Use 1¼ #8 brass screws. The screws with the smaller handles for the shutter should be fine.

Aluminum flashing — thirty feet of aluminum flashing

Waterproof glue — one pint of Resorcinol glue

This measure is based on the liquid resin. Several firms sell resorcinol glue. For availability in your area contact Roberts Consolidated Industries, City of Industry, Ca., 91749. 213:338-7311.

Wood preservative — one gallon of green #10 Cuprinol wood preservative

To our knowledge, this is the best material to use around plants. Several wood preservatives are toxic to plants, see the text on cutting. Cuprinol is produced by Darworth Company, Avon, Ct., 06001. 203:677-7721.

Paint—two gallons exterior oil based paint and one gallon primer

The green Cuprinol is hard to cover with paint. Wood treated with Cuprinol should not be painted with latex paint until it has cured for at least six months. Use a good oil based primer, and cover that with two coats of any high quality oil based paint. The inside, top frame and doors should be white to reflect light into and around the frame. However, you can paint the outside whatever color you like.

Caulk — four tubes

You will need at least three tubes of oil-based caulk, but only one tube of the more expensive silicone caulk.

Weatherstripping — 60 feet of self-sticking, closed cell vinyl foam

The foam we have used is made by Wrap-On Company, 341 Superior St., Chicago, Il., 60610. 312:822-0450. It comes in many sizes. For a properly built frame, ⅜ inch wide and 3/16 inch thick will be fine. If your doors have some warp to them, use ⅜ inch wide and ¼ inch thick foam, which will give a better seal.

Ductape—one roll two inches by 30 yards

Use aluminum ductape. This is used to tape the shutter strips to the insulation for the night shutter. There is no movement of this material, and aluminum faced tape works best. Other types of tape will work, but not as well.

In some parts of the country you may need additional copies of the blueprints to get zoning or building code permission to put up the frame. If you need builder's copies of the prints, they are available for $6.95 for the first set, and $4.95 for each additional set in that order. This is for the blueprints only, and is only offered to those who have bought Rodale's Solar Growing Frame, for their own use.

TOOLS

COMBINATION AND FRAMING SQUARE

The small square is used to mark wood for cutting, while the large square is needed to make the doors and frame square before gluing.

COPING SAW

To cut out the two latches.

STAPLE GUN

To fasten the interior glazing of polyethylene.

TIN SHEARS

One set of straight jawed shears will cut the fiberglass glazing and the flashing.

SOCKET

To fasten the countersunk anchor bolts, you must have a socket, a wrench won't reach.

POWER SAW (CIRCULAR SAW OR RADIAL ARM SAW)

One of these will be needed for the cutting of the top rail for the front and back walls. Everything else can be cut with a hand held circular saw, although either a table saw or radial saw will greatly speed up the time of the cutting, and the precision.

MISCELLANEOUS TOOLS

A 12-foot tape measure, a straight edge (a good 2 x 4 will work). Screwdriver, hammer, adjustable wrench, $7/16$ inch wrench, utility knife, wood chisel, and pencils and pen.

Safety Glasses: Because of the number of critical cuts, safety glasses should be worn, so you may watch the blade cutting, and not depend on the saw guide for accuracy.

HAND SAW

A cross cut saw will be needed to notch the plywood sides, and to cut the insulation panels.

ACID BRUSHES AND DISHES

To mix and apply the glue.

CAULK GUN

To apply the caulking to the frame.

DRILL

An electric drill and drill index is needed. All screws are $1\frac{1}{4}$ #8, so one combination bit that drills the pilot, shank and countersink can be used. A $\frac{1}{2}$ inch and $1\frac{1}{4}$ inch spade bit will also be needed.

MASONRY TOOLS

To lay up a surface bonded wall, all you need is a four foot level, a trowel and a mortar board.

CUTTING

The first step in building your growing frame is to buy and cut the wood. We've designed the project so that every piece of wood is cut first, then all wood is treated before it is assembled into five pieces and finally the pieces are put together on the foundation.

The materials list we have supplied includes enough wood to make every piece in the frame. However, that assumes you will not have any waste from split or cracked ends on wood, and that you will not make any mistakes. Depending on how far you have to travel to buy your wood, you may want to get a couple of extra 2 X 3 and 2 X 4 boards. For accuracy in the finished joints, all lumber should come from the same pallet, and be of the same thickness.

It is important that you be extremely careful during the cutting process, for any errors at this stage will show up later. By cutting all the pieces at once, the number of mistakes you make will be reduced. If you try to cut each piece as you need them, one mistake can quickly lead to another and another over the remaining cuts.

CUT PLYWOOD

The first step is to cut the plywood as shown on Sheet I. If you cut the plywood first, you will be able to use the 2 X 4 lumber to support the plywood for cutting.

After the plywood is marked, position the 2 X 4's on their wide side on the floor and lay the plywood on the boards, with the 2 X 4's about two inches on one side of the line to be cut. Place several 2 X 4's under the part of the plywood you will have your weight on while cutting. This will give you a good cutting surface, without running the risk of cutting into saw horses or a table top. With the plywood supported, be careful not to have your circular saw set too deep, or you'll cut into the floor.

If you are cutting on a table saw, you will have to eye-ball the angle cuts, which is somewhat hard to do accurately. If using a handsaw, you will have to cut the plywood on a saw horse or bench top.

The first pieces of plywood you will cut are the four sides, pieces C-6, C-7, D-6, and D-7.

Begin cutting the sides with a circular saw by placing one piece of plywood with its good side down. All the plywood should be marked for cutting on the bad side. This is important to remember. If the pieces are marked wrong, you will end up with a bad face of plywood facing outward on the finished growing frame.

On the bad side, measure and make the marks for pieces C-7 and D-6. The plywood does not have to be cut to width, it uses a full 48 inch sheet. Measure according to the dimensions on Sheet C, make the marks, using a straight edge or chalk line to mark the cut line. Do not try to mark the cut line by the given angle, use the measurements. With one cut you get the angle for two pieces. Measure piece D-6 and make this square cut. As each piece is cut, mark it with its label of the given letter and number using a ball-point pen. Repeat this process on the other sheet of plywood cutting pieces D-7 and C-6 being sure to measure piece D-7 first according to the measurements on Sheet D. Remember that all four pieces are completely different and are not interchangeable. The inside skins are longer than the outside skins, and because of the angles and the plywood being good on only one side, none can be switched. Be cautious in your measurements of the angles. Be sure each piece is measured to the dimensions given for it on its page of the plans.

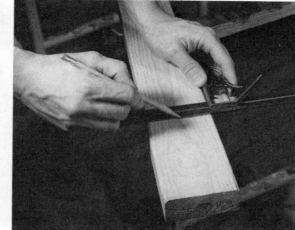

NOTCH PLYWOOD

With all four angle pieces cut, trim off the end of one full sheet of plywood to make the inside rear skin, piece B-5, and set aside an entire 48 X 96 inch piece of plywood for the outside rear skin, piece B-4.

From the plywood remaining from the angle cuts, cut the four pieces for the inside and outside front skins. Note that the inside and outside skins are different lengths and widths. With all pieces cut, take the two inside skins, pieces C-7 and D-7, and mark the notches on the bottom corners. Be sure you have the right pieces before you mark them. The inside skins are longer than the outside skins.

Using a handsaw, or saber saw, cut out the notches you have marked on each piece. Again, be sure you are cutting the notches on the inside skins, not the outside skins. Measure the pieces before cutting to be sure they agree with the piece shown on the prints.

CUT 90 DEGREE

To cut the dimensional lumber go to Sheet J, and begin at the top of the left hand column. Don't try to cut all the pieces to a page of the prints. Cut all the 2 X 8 lumber first, then move to the 2 X 4's, and on to the 2 X 3's. This way, you'll cut everything from the right stock.

To use Sheet J, read what piece you will be cutting, go to that page of the plans, look at the diagram to see what the piece looks like and its measurements. Be sure to look for any special notes on cutting steps.

As each piece of lumber is cut, use a ball-point pen to mark it with its letter and number.

Check factory cut ends for squareness before cutting your wood. If the factory cut is not square, trim it to square before measuring.

Before cutting, take your best four 2 X 3's and set them aside for pieces E-4, F-1, F-2 and F-3. Save your best two 2 X 4's for pieces A-1 and B-1. Save a good 2 X 4 for piece E-3 and the ten foot 2 X 4's for the two E-1 pieces.

CUT BEVELS

As you work your way through the cutting diagram, stop when you get to the pieces that have angle cuts. Only eight pieces have an angle. Set your saw for a 40 degree cut, and cut them all at one time.

A 40 degree cut is not interchangeable, as a 45 degree cut is. When you reverse a 45 degree cut, you get a 45 degree angle. When you reverse a 40 degree cut, you get a 50 degree angle. Be sure you are cutting the angle the proper way, especially with pieces C-1 and D-1. If you doubt the accuracy of your saw guide, cut a piece of scrap, and compare it to the angle on the plywood you've already cut. They must be the same.

If using a circular saw, you should use a saw guide, as shown above. This is a 90 degree angle made of wood or metal that when held against the wood with your other hand, gives a solid edge for the saw to be held against when cutting an angle. When cutting an angle don't watch the cutting guide, watch the blade for an accurate cut. When watching the blade wear protective goggles.

DADO CUTTING

After all the lumber is cut to length and labeled, you have to cut the half-lap joints. These are the only joints in the project, and all are cut the same.

From the pile of cut lumber, pull out all pieces labeled E-1, E-2, E-3, E-4, F-1, F-2 and F-3. There will be two pieces for each number, except E-3 which only has one piece and F-3 which has four pieces. The joints for the two E-1 pieces will be cut last, so set them aside. At this time you will be cutting joints on 13 pieces of wood.

There are two ways to cut a half-lap joint, with a dado blade in a table saw or radial arm saw, or with a circular saw and a chisel and hammer. For all the joints in this project, figure about one half hour of cutting on a table saw or radial arm saw, and about two to three hours with a circular saw and chisel. In either case, the critical measurement is the depth of the cut. The width measurement of the cut is next in importance. But, the depth of the cut is the thing most critical to a good tight joint.

All joints will be cut the same. That means that once you set your saw for one half the thickness of your 2 X 4's and 2 X 3's, it should stay the same for all cuts. Using some scrap pieces, cut a few sample joints, and be sure they fit so the tops lay perfectly flat after joining. This will be harder if done with a chisel than with a dado blade.

If cutting with a table saw and a dado blade, it is a simple matter of being sure each board is marked properly as to whether or not it will be joined to a 2 X 4 or a 2 X 3. Be sure to measure the width of your lumber before marking and cutting. You may find it best to take the actual piece of lumber that will be joined and mark its width in place for the best fit. To measure with the lumber, lay the one piece of lumber on top of the other, be sure they are square using the carpenter's square and mark the width of the joint on the board to be cut.

If using a circular saw and chisel, you are concerned with the depth of the cut as already discussed, and the placement of the markings. Run the circular saw along the two outside measurement marks once you've tested the depth, giving you the sides of the joint. Then make a saw cut about every half inch through the middle of the joint. After cutting, carefully chisel out the remaining wood, and you're set. Take your time in measuring, marking and cutting the joints, and you'll have a good tight fit when you're done.

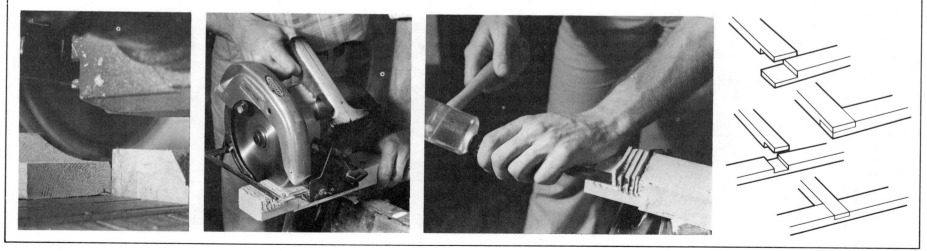

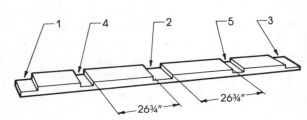

TOP FRAME

With the joints cut in all the other pieces of wood, it's time to cut the joints for the two E-1 pieces, the top and bottom rails of the top frame. Each gets five half-lap joints. One goes at each end, and is for a 2 X 4, one goes in the middle, also for a 2 X 4, and two go between the middle and the end pieces, these are for 2 X 3's. Cut the three dados for the 2 X 4's, pieces E-2 and E-3, first. Then, cut the remaining two dados for the 2 X 3's, pieces E-4.

The critical measurement is 26¾ inches from the inside edge of the center post, E-3, to the inside edge of each 2 X 3 hinge post, E-4. This opening will be for the door, and if you get it accurate at this point, the door will be much easier to weather strip.

For the best fit, put piece E-3 in place and measure 26¾ inches from both sides and make marks. Then use piece E-4 to mark the width of the joint, holding the critical 26¾ inch inside measurement. When you have marked the width of the dado, use the square to mark your cut lines.

CENTER RAIL

With all the joints cut, you are ready for the final cuts on piece E-3. Compare the piece after you have marked it to the one above. Be sure you are removing the notches from the correct face of the piece. Do not remove any of the half-lap joint.

You have to remove a ¾ inch strip from each side of the rail, the entire length of piece E-3. This can be done with a circular saw, but is much easier on a table saw or radial arm saw. When you use a table saw to cut pieces A-1 and B-1, we recommend you also cut this and both F-2 pieces.

Set the table radial arm or circular saw for an exact depth of ¾ of an inch. Using the rip guide, set it for a width of ¾ of an inch, to the inside edge of the blade, giving a finished cut width of ¾ of an inch, no more. If you measure to the wrong edge of the blade, the cut will be ⅛ inch too big. It will take two passes of the saw blade to remove each strip of wood, four passes to finish the piece of wood.

LATCH RAIL

Be aware, the two cuts on pieces F-2 described here are very easy to cut on the wrong side. Basically this is the same cut you made on piece E-3, except it's on the opposite side of the half-lap. In this case, you will be cutting a strip out of the half-lap joint, as well as the entire length of the piece. However, on this piece you will only cut one strip, not two as you did on E-3.

The photo shows how the finished piece should look after the ¾ inch strip is removed. It takes two saw cuts to remove the strip. The measurements are the same as for E-3. Both pieces should be cut at the same time, with the same setting.

Be sure you take the strip out of the right side of the wood, it is easy to get confused on this piece.

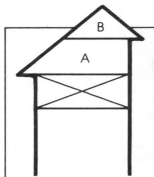

TOP RAIL

The saw cuts explained in this section are the toughest in the entire project. They also are the most potentially dangerous, so be careful, and don't try to do it yourself, have a helper.

To cut pieces A-1 and B-1, you must have either a table saw or a radial arm saw. A table saw is best. We have tried many different ways to make these pieces with a circular saw, and there is no way to do it. We redesigned the project so the piece wasn't needed, but in the end felt the advantages offered by this piece far outweigh the disadvantage of needing a table saw. If you don't have your own saw, borrow the use of one from someone. The two cuts you will be making call for one setting of the saw, and four passes through the saw. It should not take more than ten minutes. If you can't find a friend with a saw, ask around at local businesses, or try the local high school or college. A good idea is to go to your local lumber yard or mill and see if you can use their table saw.

Once you've located a saw, be sure it has a wooden rip fence. The blade will be cutting all the way through the piece of wood, and into the rip fence. A metal fence will ruin the blade as soon as they come in contact. If the saw does not have a wooden fence, you have two options, you can either put a wooden face on the metal rip fence, or clamp a 2 X 4 onto the table, and use that for your rip fence.

With the blade at the standard 90 degree angle, position the wooden rip fence 1½ inches from the inside edge of the blade. That means the side of the blade closest to the rip fence is 1½ inches from the fence. With that measurement set, turn the blade to a 40 degree angle, and turn on the saw. With the saw running, raise the blade until it just starts to cut into the rip fence. If you are using a clamped 2 X 4, the blade will miss the 2 X 4.

Begin by passing both pieces to be cut through the blade on edge. The photo shows how to position the wood for this cut. Be very careful, as the blade is cutting the wood in half, and there is nothing to steady the top piece when it is cut free. Your fingers can easily slip into the blade at the end of the cut. Have a helper pull the wood the last couple of inches, while you use a small push stick.

With both pieces through the saw, take the two smaller pieces, and placing the just cut face on the table, run them through the saw, as shown. Be sure you have the right face on the table, or you will end up with a piece cut to the wrong angle. Be careful that the sharp angled edge does not slip under the rip fence, giving you a crooked cut. This cut is not as tricky or as dangerous as the first cut, but have a helper, and use a push stick.

Of the six pieces you have now made from the two 2 X 4's, four will be used. Be sure you throw away the right piece. The finished pieces will be the top of the front and back walls. The sharp angled edges will stick out, over the plywood, forming a built-in drip edge, protecting the end grain of the plywood from moisture.

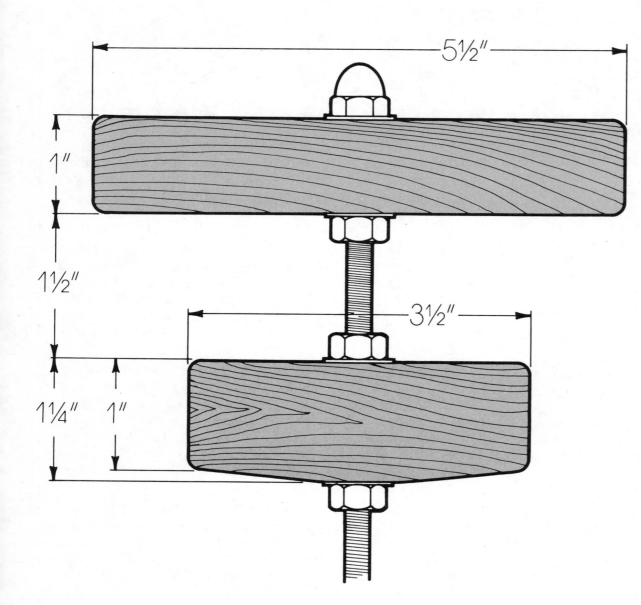

CUTTING LATCH
The solar growing frame uses two latches. These are cut from wood ¾ inch thick. If you use wood of any other thickness, the latches will interfere with closing the doors.

Each latch requires two pieces of wood, making a total of four pieces of wood for both latches in the frame. Cut the latch pieces from the end of the 1 X 6 you will be cutting the shutter strips from in the next step. If you cut the shutter strips first, you will not have wide enough wood leftover for the latch parts.

First, cut off a two-foot section of the 1 X 6 board. Remember, that is the nominal size of the wood, the actual measurement should only be ¾ inch thick. The illustration is full size, and may be used as a cutting diagram for the latch parts. Use a piece of carbon paper and trace the pieces, then transfer them to the wood and cut them out. Be sure you have the grain running in the right direction, or the wings will easily break off. After the pieces are cut, drill a ¼ inch hole through the center of each piece. Don't forget, you need two latches, a total of four pieces.

SHUTTER STRIPS
After cutting the latch pieces from the 1 X 6, you have to cut the strips that will be used to strengthen the night shutter. All you have to do is rip four strips out of the 1 X 6. Each strip should be exactly 1 inch wide so they will perfectly match the 1 inch thick insulation board we recommend for the shutter. If you are using a night insulation of a different thickness, rip the strips to the exact thickness of the insulating board. A table or radial arm saw is best for this operation, although a circular saw or even a sabre saw will work. If you are careful, even a hand saw will work just fine. You'll be cutting these pieces to their exact length when you install the shutter. For now, make them 8 feet long.

CUPRINOL

With all the wood cut and the joints cut, it's time to treat everything with a wood preservative. As discussed in the materials section, we use Cuprinol to treat the wood. Every surface of every piece of wood should be treated.

Use piece B-4 as your work table for treating the wood and all spilled material will end up treating the rear wall, and nothing will be wasted.

For the absolute best protection, the end grain of the wood should be soaked in preservative for a few minutes. You can do that by standing the wood on end in a bucket of preservative. If you soak the end grain, you may need more than one gallon of preservative. If you don't soak the ends, one gallon will be enough for the entire project.

As you treat the wood, stack it by the letter, so all pieces for each page of instructions will be together. This will speedup assembly.

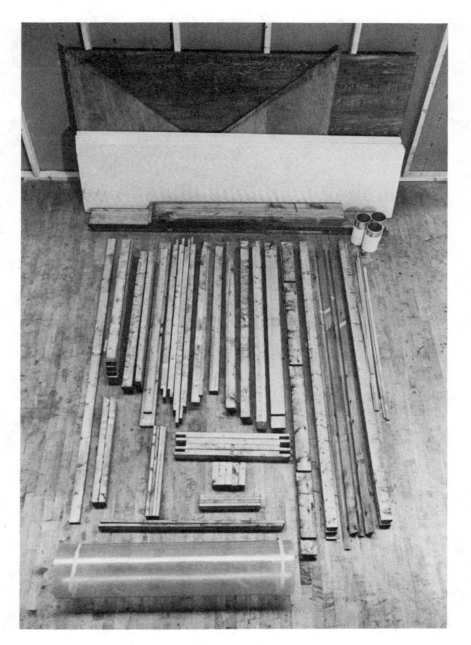

FOUNDATION

Of the entire growing frame construction process, this may be the area that intimidates the most people. There is something about the word masonry that sends shivers down the spine of do-it-yourselfers. With that in mind, we have used a foundation technique that does not require traditional masonry skills. We use surface bonding cement for the foundation instead of standard mortar. The concrete blocks are put in place dry. When they are in perfect position, even if it takes you all day to readjust them, a coating of the stucco like cement is troweled on the outside of both walls, bonding the blocks together.

Surface bonding cement has been found to be stronger than regular mortar in government tests. Fiberglass strands in the cement give the wall more resistance to side forces than a regular mortar wall.

If you know how to lay up conventional block, put your foundation in that way, it's cheaper. But if you don't know how to lay block, and still want a permanent foundation, do it as described here. If you don't want any foundation, see the end of this section, for how to build an above ground growing frame.

The first step in putting in a foundation is to dig a hole. The measurement of the finished foundation should be approximately 55 inches X 102 inches.

If you live in an area that does not receive deep frost, you will not need any special foundation insulation. However, if you live in an area that gets ground freezing you will either have to put in a footer, a layer of concrete, below the frost level to support the frame, or use the tundra insulation technique we show. This uses insulation extending outward from the foundation to prevent frost from getting under the foundation and heaving it.

A foundation using tundra insulation needs a hole two feet larger on all sides, giving you a rough opening of almost seven feet by thirteen feet. The hole should be 26 inches deep. Using three layers of block, this will put the finished foundation just above ground level.

If you are not using tundra insulation, the hole you excavate need not be as large, but to put in a

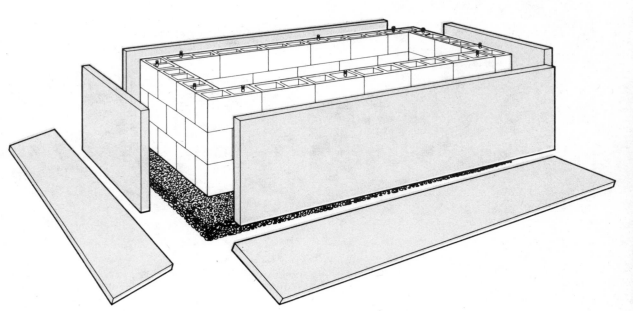

footer you will have to dig at least 8 inches below your frost line.

The illustration shows how to mark the hole with string, so a backhoe won't destroy the string lines while digging. Take care to mark the hole off accurately, and it will help you lay up the block walls properly. Don't remove the string after the hole is dug. You will need it to get the first course of block straight.

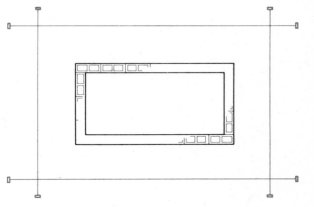

GRAVEL

After the hole is dug, fill it with 4 to 6 inches of fine gravel. A gravel with an average size of no larger than ½ inch is best. Rake the gravel to an approximate level, and then level it exactly with a 2 X 4 and level, as shown in the photo.

With the gravel smoothed to a level base, walk over the area where the blocks will go several times to compress the gravel. An asphalt tamper does this job best, but walking on the area where the blocks are to go will also work. After walking the gravel flat, check it again with the 2 X 4 and level, and fill in any low spots.

BLOCKS

With surface bonding cement, the most critical aspect of the wall is the first course. If that first course is square, plumb and level, you will have little trouble with the rest of the wall. Take your time laying the first course, and check all measurements to be sure everything is right.

Begin positioning blocks at a corner, making sure both walls are the proper distance from the edge of the hole. To test this, hold a level against the block, as if checking for plumbness and measure from the side of the level to the string that you put up to mark the foundation outline. The illustration shows how to check for this measurement.

Continue to lay the first course, checking along the tops of the blocks for level, and along the sides to be sure they stay flush. Also check the distance from the string and use the carpenter's square to get the corners in line. Measure from one wall to its parallel wall to be sure they are parallel when you start the other walls.

After the first course is completely in place, and you have a rectangle of blocks check from the front to back wall and across corners for level. Occasionally check various blocks to see if they are plumb, and check from corner to corner with a tape measure to see if the unit is square. The page on checking for levelness shows all ways to check for proper alignment of the blocks. Be sure everything in the first course lines up exactly.

With the first course in place and checked all ways for alignment, begin putting on the second and third courses. Don't put the blocks exactly on top of the first course, offset the corners for greater wall strength as shown in the drawings.

You will find that some blocks do not sit squarely on other blocks. For wobbly blocks, use small stones as shims to get the blocks in line and firm. After the third course is in place, begin all the level and squareness checks again. Continue to adjust the blocks until you are sure everything is straight, plumb and square. Double check one more time before getting ready to put the cement on the blocks.

LEVELING

There are many ways to check alignment of the blocks. The beauty of a surface bonding cement is that you have ample time to check each block as many ways as you want, there is no rapidly hardening mortar to speed you along, often forcing errors from beginners.

The first way to check blocks is for levelness. This is done by laying a level across the top of the blocks. Take more than one reading, one on the outside edge of the foundation and one on the inside. Before reading the level, work it back and forth on the blocks several times to smooth out any small imperfections. Every block in the first course should be checked this way. If it is not level, either add a few stones under a low edge, or remove some from a high edge.

Also check level from one wall to another. This is best done with a four foot level. If you are working with a two or three foot level, use a straight length of 2 X 4 to get a reading from one wall to the other. The illustration shows the ways to check from wall to wall with a four foot level.

Also check the flushness of the wall. Use the side of your level, or a straight 2 X 4 for this. Place it across the outside face of the blocks, and looking straight down from the top, check that all blocks touch the level. Any blocks that are canted inward or outward will quickly show up. Just tap them back in place with your hand, until that section of wall is lined up.

Check the alignment with the string while doing this. It is best to check the first and last block of a section for alignment with the string first, and then check that same section for flushness.

With the first course in place, check the unit for squareness. This should first be done with a carpenter's square, when the individual corners are formed. Once the finished rectangle is laid out, it should be checked for overall squareness. To do this, measure the distance from one corner to the other, diagonally across the frame. Do this both ways. If the frame is square, the measurements should not be more than ¼ inch different. If they are off by more than ¼ inch, you've built yourself a parallelogram. Use the carpenter's square and straighten the two side walls first, and then the back wall, after first checking to see that the front wall is in line with the string.

After being sure the first course is perfectly aligned, lay up the next two courses, being careful to keep them lined up as you do. When all three courses are in place, check the wall for levelness and squareness.

With everything checked, and double checked, you are ready to apply the cement.

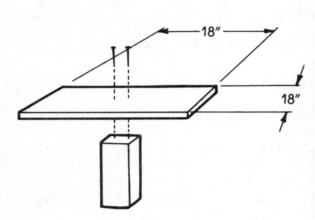

18"

18"

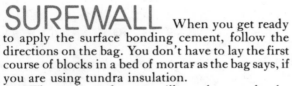

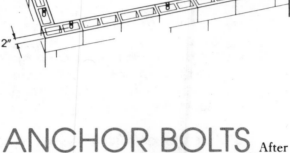

2"

SUREWALL
When you get ready to apply the surface bonding cement, follow the directions on the bag. You don't have to lay the first course of blocks in a bed of mortar as the bag says, if you are using tundra insulation.

The only tools you will need to apply the cement are a square trowel and a hawk (a platform to hold mortar or cement while you put it on the trowel). An inexpensive hawk can be made from a piece of plywood, with a handle on one side to hold it upright. The drawing shows a plan for making a hawk that will serve you well.

When mixing the cement, be sure to follow the water recommendations exactly as they appear on the bag. Too much water will make a mix that is too thin to stick to the blocks, and will not give you a strong wall. Too little water, and the material will not properly spread and you will use too much cement.

Don't forget to wet down the blocks before putting on the cement. Coat the outside walls first, working from the bottom of the blocks up to the top, pulling the cement in arching strokes towards your body. As you cover an area, move backwards, again, pulling the cement across the blocks towards your body, as shown.

Don't try to get the cement overly smooth, it will be covered by soil. Do try to get an even coating for increased strength, and keep the corners as neat as you can. Smooth out any lumps on the surface of the blocks, so the insulation will fit well.

After doing the outside walls, move to the inside. Don't press too hard on the walls when putting on the cement, as they can be pushed out of line fairly easily at this time. Also be careful when stepping over the wall, your trailing foot can easily knock a block or two out of line. This is a time to step lightly around the foundation. After the cement is applied to all walls, remember to wet down the walls several times as recommended on the cement bag.

One last point, be sure to wash off your tools completely between each batch of cement you mix. If you leave cement from the last batch on your tools, or mixing container, it will greatly speed up the setting time of the cement. This will make it much harder, if not impossible to apply the entire batch smoothly. After doing one batch of cement wash up your tools, take a break and check out your progress. After resting, start again on a new batch, with clean tools.

ANCHOR BOLTS
After the cement is on the blocks mix a bag of standard mortar cement. Add a lot of small gravel to stretch the cement.

Look at the diagram to see where the anchor bolts are to go. Wherever a bolt is to go, stuff rolled up newspapers down the cavities of those blocks. This reduces the amount of cement you will need, and does not hurt the strength of the bolt. Be sure you keep the level of newspapers about 3 inches below the bottom of the top course of block. If the bottom of the anchor bolt is in the papers and not the cement, it will not hold.

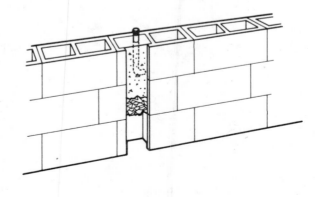

Put the anchor bolts more towards the outside of each cavity than the inside. The sill plates will be offset from the inside of the foundation opening somewhat, and putting the anchors towards the outside will give more strength to the unit.

The bolts should be no more than 1½ inches above the level of the blocks. Fill the block cavity with mortar, tap it down, smooth it out, and then push in the anchor bolt. Using the side of your trowel as a base to measure from, be sure the bolt is exactly 1½ inches above the blocks and no more.

Put in all ten bolts, making sure they are the right height and set towards the outside of the openings, and your foundation is done. In a couple of days you can begin assembling the growing frame.

FREESTANDING

We have been experimenting with freestanding solar growing frames. These units are built the same as those with a foundation, except the walls are elongated, to make additional room for soil. The units are either set on top of the ground, or dug into the ground.

At this time we know they are cheaper to build, and if protected from winter winds with hay bales or leaves, they should perform as well thermally as those units with a foundation. The thing we don't know about these units is how long they will last. With the wood in constant contact with the soil, we don't yet know what type of life expectancy these freestanding units will have. One plus, the most difficult part of the frame to build, the top and doors, will not be hurt by the soil, so the worse you might have to do is rebuild the walls every four or five years, depending on how well they hold up to the moisture.

With the unit above ground, the doors will normally be above snow level, reducing the need for shoveling. In places with extremely hard soil, or rocky soil, an in-ground foundation may prove to be impossible. Plus, for those who just don't want to dig a foundation hole, an above ground installation makes a solar growing frame a reality.

As discussed earlier, the key to the solar growing frame is the insulated soil bed. At first glance, you might think building the frame above ground would sacrifice the warming effect of the earth, it doesn't. Earth is not a good insulator, it is a good moderator of temperature extremes. An above ground frame can receive this same benefit from stacked up bales of hay, bags of leaves, or even cornstalks. The primary insulation, the foam in the walls and under the soil bed, and the night shutter, will be enough to protect the frame from cold weather in most locations.

In an urban setting, installing the above ground unit on a flat rooftop would be one way to grow your own food. To put a frame on a rooftop, first check to see that the frame will not interfere with water runoff on the roof, or create undue snow drifting. Also check to see that the roof can support the weight of a growing frame. The wooden structure of the frame weighs about 200 pounds, and the 12 inch soil bed weighs about 4,000 pounds. This load can be supported by most roofs, especially if you place the frame over load bearing walls or supports. However, you must check with the contractor or designer of the building to make sure the roof can safely support the frame.

To build an above ground growing frame, use the same plans as for an in-ground unit, with the exceptions noted on the next two pages. You will need to buy more wood, but no masonry materials and less insulation. The hardware stays the same, except you need additional screws, nails and chain.

The only major changes you will be making in the plans to make a freestanding frame is to make the four walls taller, strengthen the assembly of the walls, add an additional support and pay more attention to waterproofing the walls. Besides that, an above ground frame is no harder to build than an in-ground frame.

The first changes you will make involve going over the plans, page by page and making all the measurement changes listed on the chart. Be sure you don't miss any pieces. Any changes should also be marked on the cutting diagram, Sheet J, for the dimensional lumber. Any pieces that don't have changes should be cut the same as before. Measure carefully for good economy of materials on any pieces that have changes.

Completely disregard the plywood cutting and insulation cutting diagrams on Sheet I. Use the cutting diagrams in this section for the plywood and insulation. Add to the materials list those items listed in this section although you should still read the other materials list to get a better understanding of what materials you need, and why.

After cutting all the pieces, disregard the foundation instructions completely. When putting the walls together, the only changes are that there is no plywood overhang at the bottom of the inside wall.

The other change in wall construction is that you should put the outside skins on the wall during the construction phase.

When nailing the inside rear skin to the wall, be sure to put the larger of the two pieces of plywood on the bottom, so the seam where the two pieces join is not covered with dirt. Any seam below ground level will be an invitation to moisture.

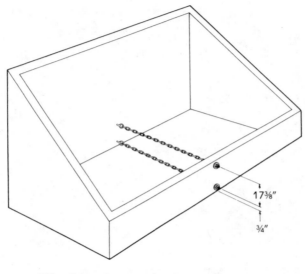

The shutter support strip measurements will change. Recalculate these measurements by figuring the front height the same way, and add 9 inches to the rear wall height given in the other section. With the four walls built, put the top and doors together exactly the same as in the plans, there are no changes to be made in this section at all.

When you get ready to assemble the unit on the site, instead of having a foundation, you will need a gravel pad, about 4 inches thick, and about one foot bigger than the outside dimensions of the frame in all directions. The frame just sits on this gravel pad.

To put the four walls together, use 1 inch screws instead of nails to fasten the overhanging plywood of the front and rear walls to the side walls. Round head brass screws with washers will give the most strength. Follow the same cautions to be sure everything is lined up.

With the walls firmly erected, install the center chain supports. These are needed to give the unit added strength to hold the soil inside the frame. In the ground, there is soil on both sides of the walls, and the stress is not as great. However, the above ground unit needs the added strength of the chains.

To make the chain supports, use four eyebolts, four inches long. Drill a hole through the center stud in the front and rear wall, and insert the eyebolt through it. Do not just put the eyebolts through the walls, they must be through a stud.

Install the chain, and tighten the nuts from the outside, until the chain is tight. Don't put too much pressure on the chain, it will pull the front and back walls out of line. Just get it snug, the goal is to have it help the walls resist the pressure of the soil.

Painting

Paint the entire unit, before putting on the top. After painting, coat the inside of the frame with an aluminum fibered roof coating to give the wood one last layer of protection from moisture. Put the roof coating on heavily, being sure it seals well in the corners, and on the bottom. Cover the bottom 14 inches of the frame for a 12 inch soil bed.

We have found that an aluminum fibered coating will not harm plants once it has dried. However, a great many other roof coatings will. Any material that is arsenic based or includes creosote will retard plant growth, stay clear of them.

With the inside painted, install the shutter and the top, being sure to weatherstrip all cracks, and your above ground unit is set to go. Garden it the same as an in-ground unit, and you should have no problems.

Watch it carefully over the years for signs of deterioration of the wood. We won't know until our units begin to show signs of deterioration how long they will last.

One last option is to build a freestanding unit, and bury it part way in the ground. To do this, coat both sides with aluminum roof coating before putting it in the ground. Dig the hole to the depth you want, and put four inches of gravel on the bottom. Be sure to put in bottom insulation on any freestanding growing frame.

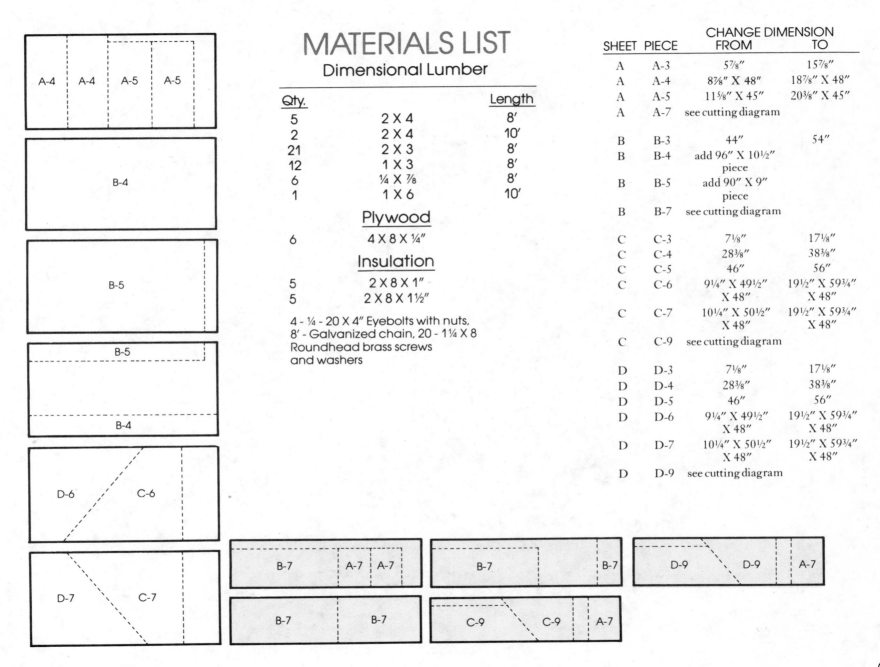

MATERIALS LIST
Dimensional Lumber

Qty.		Length
5	2 X 4	8'
2	2 X 4	10'
21	2 X 3	8'
12	1 X 3	8'
6	¼ X ⅞	8'
1	1 X 6	10'

Plywood

6	4 X 8 X ¼"

Insulation

5	2 X 8 X 1"
5	2 X 8 X 1½"

4 - ¼ - 20 X 4" Eyebolts with nuts,
8' - Galvanized chain, 20 - 1¼ X 8
Roundhead brass screws
and washers

SHEET	PIECE	CHANGE DIMENSION FROM	TO
A	A-3	5⅞"	15⅞"
A	A-4	8⅞" X 48"	18⅞" X 48"
A	A-5	11⅝" X 45"	20⅜" X 45"
A	A-7	see cutting diagram	
B	B-3	44"	54"
B	B-4	add 96" X 10½" piece	
B	B-5	add 90" X 9" piece	
B	B-7	see cutting diagram	
C	C-3	7⅛"	17⅛"
C	C-4	28⅜"	38⅜"
C	C-5	46"	56"
C	C-6	9¼" X 49½" X 48"	19½" X 59¾" X 48"
C	C-7	10¼" X 50½" X 48"	19½" X 59¾" X 48"
C	C-9	see cutting diagram	
D	D-3	7⅛"	17⅛"
D	D-4	28⅜"	38⅜"
D	D-5	46"	56"
D	D-6	9¼" X 49½" X 48"	19½" X 59¾" X 48"
D	D-7	10¼" X 50½" X 48"	19½" X 59¾" X 48"
D	D-9	see cutting diagram	

More than anywhere else in the project, here is where you will see how accurately you cut the wood. Even minor mistakes in cutting will show up when you start to put the top and doors together.

This unit is made up of one top frame, and two doors. First they have to be put together as separate pieces, then the pieces put together to make the unit.

The first step is to assemble the top frame. Lay out pieces E-1 (2), E-2 (2), E-3 and E-4 (2) on a flat surface. A level garage or basement floor is ideal. The top will be assembled by gluing the top and bottom rails to the two end posts first, and then putting in the middle three pieces. Each joint should be glued, and two screws put in to give added strength.

The challenge is to get the frame perfectly square. First, lay out the frame and put all the pieces in place, without glue on them. The frame is designed so the joints are on the bottom when it is installed, but when you are working on it, the joints should face up. Check to be sure every piece fits properly, and no trimming is needed. Using a carpenter's square start at one corner, and square the end post and top rail. Go to the other corner of

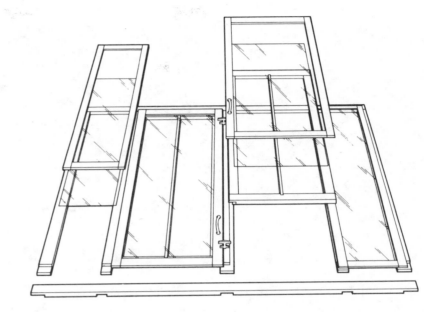

that same end, and align that corner. Go to the other end and align those corners. Recheck the corners and adjust until all are square.

When you have the top square with the carpenter's square, use a tape and measure from corner to corner (as explained in the Foundation section) to check for squareness. The measurements from corner to corner should not be more than ¼ inch different. If the difference is more than ¼ inch, you have a parallelogram. Go to a corner and start squaring the frame again.

With the frame perfectly square, gently remove the two end rails. Mix the glue, and apply it to all sides of each joint. On the end joints, that means glue on three faces of the wood.

Put glue on both end posts and the end joints of the top and bottom rails. Using a finishing nail, tack the frame together by putting one nail at each end joint. Check with your tape measure to be sure it is square, drill a hole in each end joint and put in a screw. Put only one screw in each corner first, recheck to be sure it is square, and then put in the

second screw for each corner.

With the end posts in place, glue and screw the three center pieces to the top and bottom rail. Use two screws at each joint for these pieces.

The doors are put together the same way as the top frame. Each joint is glued, and fastened with two screws. Check that all pieces of the door fit, before gluing. Glue two pieces together to form an L, and screw the joint with two screws. Do the same with the other two pieces of the door. Then, glue the two L's together, checking first with your tape measure to see that the door will be square. If you put the two L's together carefully, using the square to keep the pieces in line, the two L's should go together with no problem.

With the door glued and screwed together, glue in the glazing support strip F-8 using a finishing nail to hold it in place at each end.

With both doors assembled, the top and doors are ready for painting. Be sure to also paint all the batten strips now, both the interior and exterior battens.

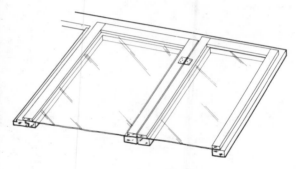

GLAZING - DOORS

With an assembled and painted door in front of you, with the smooth outside facing up, put on the glazing. The measurements given for the glazing material should be about 1½ inches bigger than the inside of the door on all sides, but check before cutting. Put ⅜ inch bead of silicone caulking around the door frame about ½ inch from the inside edge. Lay the glazing material on the silicone. Leave about ¼ inch from the glazing to the outside edge of the door frame on all sides.

Put a vertical batten strip, F-4, flush with the outside edge of the hinge rail side of the door. Be sure to check that the piece is flush with the hinge side of the door, and centered, with about ¾ inch overhang at the top and bottom of the door. Drill holes through the batten, the glazing and into the door frame with a combination bit for 1¼ inch X 8 screws. Drill one hole at a time, working from one end to the other, putting in a screw every 8 inches. To help keep the screws in line, first make a light pencil mark 1 inch from the outside edge of the batten, and put all screws along this line.

The other battens are installed the same way. The only difference is that the top and bottom battens must have ¾ of an inch overhanging the outside of the door frame, and the vertical batten on the latch side of the door must have only ½ inch overhanging the outside of the door frame. These overhangs are very critical, and must be closely maintained.

Caution: The glazing material has an inside and outside face. When the material comes to you, these will be marked. Put a piece of tape on several areas of the Kalwall to mark the outside face. Whenever you cut a piece, be sure a piece of tape marks the outside surface, so it can be properly installed.

FRAME GLAZING

To apply the outside glazing to the top, follow basically the same procedure as putting the glazing on the doors.

The pieces of Kalwall should be about 1½ inches bigger than the opening in all directions. Put the battens that cover the hinge rail on first, keeping them flush with the side of the hinge rail nearest the door opening. This gives the door a flush surface to butt against, and improves the weatherproofing of the unit.

Apply a bead of silicone caulk all around the opening, and drill and fasten the batten strip with screws every 8 inches. After the batten strip on the hinge rail side is in place, put on the other three batten strips, making sure they are lined up flush with the inside of the glazing opening. These batten strips should not overlap the outside of the top, as the strips on the doors did. Keep the inside edge of the batten strip flush with the inside opening of the frame.

DOORS
With the glazing and batten strips on the doors and the top, you are ready to fasten the doors to the frame.

The doors are designed to have ⅛ inch of play on one side, and ¼ inch of play on the other three sides. This allows the wood to swell during the winter yet still open and close. In some climates, you may still need to shave additional material off the doors if swelling is a problem. The most critical measurement is along the hinge rail side. This side should be as close to ⅛ inch as you can make it. We recommend you cut several shims off a piece of scrap 2 X 3, exactly ⅛ inch thick. Place these shims along the frame, and put the door in place. Be sure to have the door lying flat and that the two batten strips the hinges will be screwed into are aligned.

Position the handles in the center of the door, and screw them to the batten strip. Unless the handles come with very small screws, use the ones provided. If you use 1¼ inch X 8 screws be sure the screws go into the door frame, not just the batten strip.

If the door is out of line now, you will have to try to rehang it or cut it down later. Take a few extra minutes of care and get it properly centered. Once centered, screw the hinge to the top frame first, and then to the door using the 1¼ X 8 screws, not those that came with the hinges. Put one screw in each hinge, and give the alignment a final check before putting in the rest of the screws.

WEATHER STRIP
To have an efficient frame, it is important that the doors be well weather stripped, and tightly sealed. To weather strip the doors, use ¼ inch thick closed cell adhesive weather stripping on the underside of the batten strips, and ³/₁₆ inch foam on the side of the hinge rail.

The illustrations show where to position the weather stripping on the doors. It is important that you roll the material from one end to the other, fastening it as you go. If you work from both ends, you can easily end up with a bubble in the middle.

We have put the weather stripping on doors instead of the frame, to reduce the chance of it being damaged when you use the frame. There is very little stress on the weather stripping in this position, and it should hold up well. If it does come loose, reattach it with any commercially available contact cement.

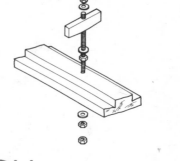

LATCH
This is a two purpose latch. It will keep the doors tightly shut, or hold them in a fixed, open position for venting. The frame uses two latches.

To the pieces you have already cut, insert a ¼ inch brass rod through the hole in the larger of the two pieces that make up the latch. Put a washer and ¼ X 20 cap nut on the end of the rod. Insert a washer and a hex nut on the bottom of the piece, and tighten both nuts tight.

Insert another hex nut and washer and slip the smaller piece of wood on the threaded rod. Be sure the flat edge of this piece is facing the edge of the piece already tightened. Allow 1½ inches of space between the two pieces of wood by adjusting the hex nut and washer placement. When it is on right, put another washer and hex nut on the rod, tightening it against the bottom of the smaller piece of wood.

To position the latches, measure about 12 inches from both the top and bottom edge of the door, and make a mark on the center rail. Be sure this mark is exactly centered between the two doors. Drill a ¼ inch hole on the mark.

Insert the rod through the hole from the bottom, and put on a washer and a nut. With the doors closed, adjust the bottom nut until there is a firm fit. Attach a second nut and tighten it up against the first nut to lock the two in place. Cut off the extra threaded rod. Repeat the procedure for the other latch.

WALLS

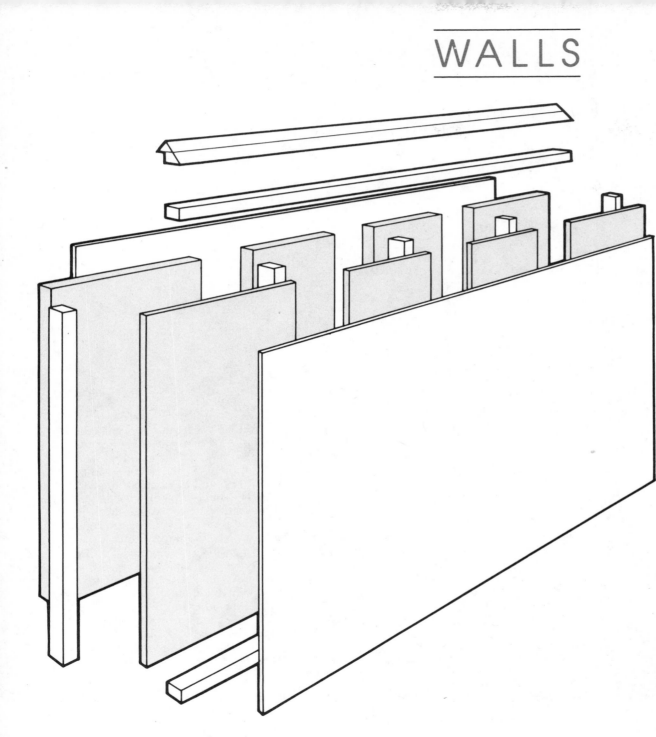

Putting the walls together is the easiest part of building a solar growing frame. The walls simply use conventional stud wall construction techniques. All pieces are butted together, no joints.

There is literally almost nowhere you can make a mistake, if you originally cut the pieces right. The only mistake you may make is to nail the outside skin on the wall. The outside skins are not nailed to the walls until the assembly phase, much later. For now, pick out all the outside skin pieces, A-4, B-4, C-6, and D-6, and set them aside.

Begin by putting the front and rear walls together first, then put the side walls together. We will explain how to put the rear wall together, the front wall is done exactly the same. Then we will explain how to put the right side wall together, the left is done the same.

Start by locating the bottom and top plates, B-2, and three studs, B-3. Nail the two end studs and the center stud in place on the bottom plate. Be sure the center stud is exactly centered, and the end studs are at the very end of the bottom plate. Use two nails to hold each stud in place. If only one nail is used, the stud will be free to twist, putting the wall slightly out of line. With these three studs in place, nail on the top plate.

A hint to keep in mind is to dull the points on the nails before nailing. By lightly hitting the point of the nail with the head of your hammer, the chances of the wood splitting will be greatly reduced. This is especially valuable when putting a nail close to the end of a piece of wood, where cracking is a good possibility.

TOP RAIL

To put together the two pieces of the top rail begin by nailing part A of piece B-1 to the top plate of the wall. When nailing this, be sure to have the angle and the overhang properly positioned. The critical measurement is that the 90 degree angle be flush with the edge of the top plate as shown in the detail on the plans. With this flush, and the ends aligned, nail part A in place with 12d nails.

Before putting part B on part A, mix up some resorcinol glue. Mix enough glue at this time to also put on the shutter support strips.

Check to be sure you have the right faces of the wood going together before applying the glue. Apply the glue to both parts A and B, and put part B on top of part A, making sure the angle is the same. When putting these two pieces together, the critical measurement is that the angle stays the same, the amount of overhang is not critical. Use 4d nails to fasten part B to part A.

Next, nail the inside skins to the front and rear walls with the barbed nails. Keep the skins flush against the overhang of the top rail.

SIDES

The side walls are put together basically the same as the front and back walls. To explain the side wall assembly, we'll go over the details of putting the right side wall together, the left side is done the same way.

Nail pieces D-3 and D-5 to the bottom rail, D-2. This is best done with all pieces on their sides, on a level floor. Put your knee or foot on the stud, and nail the bottom plate to it. Be sure the studs are lined up with the end of the rail before nailing. Now, nail the top rail to the two end studs, again, making sure the ends are aligned. Also check to make sure you have the alignment correct, compare it to the illustration on the plans, Sheet D.

Because of the angles involved, the side walls may not be too solid at this point, don't worry, once the inside skin is nailed on, it will tighten up.

To fasten the inside skin to the side wall, lay the wall on the floor, and set the skin on top of it. Working from point 1 in the illustration above, align the plywood with the corner of the top rail, D-1, and the rear post, D-5.

Tap the top rail onto the rear post to be sure it is tightly in place, and sink a barbed nail into the plywood and the top rail. Put another nail into the plywood and the rear post. Don't hit these nails home, leave about ½ inch showing so they can be pulled, if the walls are found to be out of line.

Next go to point 2 and get the plywood in line with the top rail and the front post. When these are in line, put nails in the same way, one in the top rail, one in the front post. Making one last check that everything is in line, put nails in at point 3, hitting the rear post and the bottom plate. Don't forget that the plywood should overlap the bottom plate by 1 inch. Lastly go to point 4 and put two nails in, one in the front post and the other in the bottom plate.

Make one last check to see that the plywood is in perfect alignment with the studs all around, and hit all the nails home. Put in additional nails, about every eight inches. Don't forget to nail the skin to the center stud, D-4.

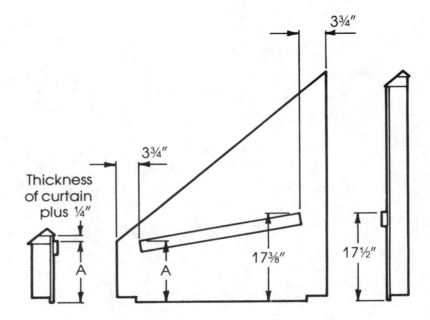

INSULATION

To install the insulation in the walls, two pieces are used at the factory 24 inch width, and two must be cut to fit. Using a sharp utility knife, or a saw, cut one full width piece of insulation to the length of the inner wall opening. Place this piece against the end stud, and pressing the middle stud against it, nail the middle stud in place. Putting the insulation in first and then nailing the stud in place gives a much tighter seal than if you were to try and nail the stud in place with an exact 24 inch opening for the insulation.

With the stud in place, repeat this process for the other end of the wall. When this is done you should have a studded wall, with insulation in two cavities, and two uninsulated cavities. Cut the insulation to fit the other two cavities, paying attention to the cutting diagram for economical use of the foam insulation.

Remember that you need two sets of insulation for each wall, one from 1 inch and one from 1½ inch foam. Cut the second set of pieces at this time. The front and rear walls are not yet complete, but set them aside for now and begin work on the side walls.

SHUTTER STRIP

The exact positioning of the shutter support depends on what thickness insulation you use. We do not recommend anything less than ¾ inch for the shutter, with 1 inch being recommended. The figures given here and in the plans are for a 1 inch shutter.

To calculate the proper positioning of the four shutter support strips, A-6, B-6, C-8, and D-8, begin with the front wall. Add ¼ inch to the thickness of the insulation to get the distance from the top of the support strip to the bottom of the overhang of the top rail, A-1. Mark this distance across the front wall, and nail the support strip in place. Be sure the nails go into studs, not just plywood and insulation.

On the back wall measure 17½ inches from the bottom of the plywood, and mark a line for the top of the support strip, B-6. Be sure you nail into studs, not just plywood.

To position the support strip on the side walls, mark the positioning for the rear corner first. Measure 3¾ inches in from the edge and 17⅜ inches up from the bottom of the plywood, not the cut out portion of the plywood. Make a mark at this point for the top edge of the strip.

On the front of the side wall measure in from the edge 3¾ inches. Measure up from the bottom of the plywood 10¼ inches. Make a mark for the front top corner of the support strip.

Apply glue to one side of the support strip and the wall. Position the strip, and nail it in place. The only stud you will be able to hit is the center stud. Nail the two ends into the plywood, to help the glue take hold.

If you are using insulation thinner or thicker than 1 inch, compensate by following this procedure. Set the front and back support strip as already explained. Put the rear side support marks in the same position, and subtract ⅛ inch from the measurement used for the front wall for the front position of the side strip.

If the finished shutter sags, tape another pull tab at the front and lift the shutter in place, or nail an additional furring strip onto the front support rail to catch the sagging shutter.

ASSEMBLY

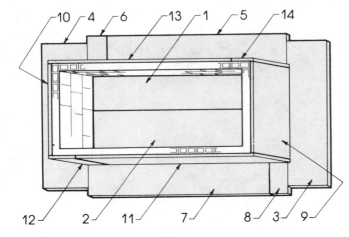

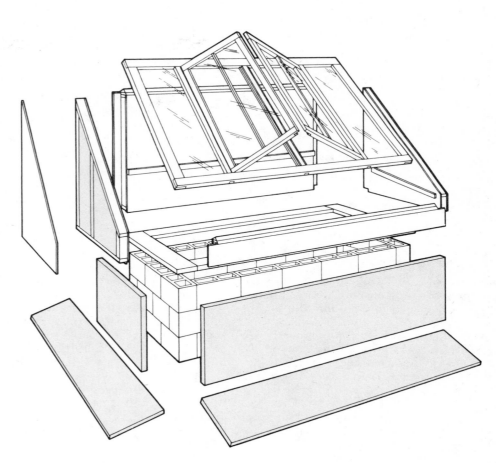

The end is in sight. You've cut all the wood, put the pieces together into four walls and a top. You've excavated a hole and built the foundation. All that remains to be done is put up the four walls, fill it with soil, put on the top and start gardening.

If you were careful in all the other steps, assembly should go smoothly. Take your time putting the unit together, and check everything pointed out in the instructions. It is only human to start hurrying near the end of a project, only to get a piece or two out of line, to find later that the top won't fit, or will fit, but not tightly and will have numerous air leaks. To be effective, the unit has to be assembled tightly. Infiltration of cold air will greatly hurt its performance.

The first step in putting the unit together is to insulate the foundation. Follow the cutting diagram on Sheet G to see how to cut the insulation. The two pieces of insulation for the inside bottom of the frame's soil box are not shown on that diagram. Cut these two pieces to length, and trim one to width for a tight fit on all sides.

Next, begin putting in the tundra insulation. These sheets of insulation should have a slight slope to them, away from the frame, directing any runoff water away from the soil box.

With the tundra insulation in, put the insulation against the walls, use backfilled dirt to hold them in place. A layer of gravel around the joining edge will also help drainage. On a three block high foundation, a 24 inch piece of insulation will have to be cut. Lean the insulation in place at the bottom, and mark a line across the top of the cinder blocks. This piece should be cut off, and the insulation will fit perfectly.

If you have some other type of foundation, the critical point of the foundation insulation is that it end up flush against the bottom sill plate. That means it has to be level with the top of the concrete blocks. If you have more than 24 inches of space between the tundra insulation and the top of the blocks, the tundra insulation has to be removed, the space filled with dirt or gravel up to the 24 inch level, and the insulation replaced. The finished insulating job should have the foundation insulation extending from the top of the blocks down to and touching the edge of the tundra insulation.

The diagram shows the order in which to put all the insulation pieces for the foundation. Follow it and you should have no problem.

Figure labels (left diagram):

A

B

Location of hole for anchor bolt

½ of (90½ minus A)

Distance from inside face of block to center of anchor bolt minus ½ of (43 minus B)

Mark indicating setback of end sill plates, lined up with inside edge of block

(middle diagram):

Side setback

Front setback

SILL PLATES

If you measure the inside opening of the foundation, you will notice that it is not wide enough for the walls to fit. The sill plates, pieces G-5 and G-6, must be set back from the inside face of the blocks to make a proper inside opening. The sill plates should have an inside dimension of 43 inches across and 90½ inches long.

To calculate how much of a set back the sill plate needs, first take measurement A, subtract that from 90½ inches and divide the sum by two. This gives you the amount of set back for each end sill plate, G-6. Using this measurement, mark it on each end of the front sill plate, G-5 as shown in the illustration. Line these marks up with the side walls as shown.

With the marks lined up, mark the center of each anchor bolt on the edge of the sill plate. Using the combination square score a line across the top of the board to give the anchor bolt location.

Next, subtract the actual width of the foundation opening, measurement B, from 43 inches and divide that sum by two. This will give you the needed set back for the front sill plate. Measure the distance from the inside face of the block to the center of each anchor bolt. Subtract the calculated set back, and mark that distance on the anchor bolt positioning lines you have already marked on the board. This gives you the exact center of the anchor bolt, for the needed set back of the sill.

With the 1¼ inch spade bit, drill the countersink hole. Go deep enough for an anchor bolt nut and washer. After the countersink, drill the ⅝ inch hole for the bolt. Don't drill the small hole first. Put the sill plate in place, and lightly tighten the nuts.

With the front sill in place, move to the end plate. You don't have to go to all that trouble for every sill plate, only the first one. If the first one is in properly, the rest will easily follow. If the first one is in wrong, it is hard to compensate with the others. The front sill already gives you the positioning of the end plate, all you have to do is keep it square, mark the anchor bolt position lines, and calculate the set back for the proper center of the anchor bolts. With that done, drill and fasten the end plate in position. Do the same for the rear plate and the other end plate. Before tightening all sill plates firmly down, check the squareness of the whole unit by measuring diagonally from corner to corner as shown in the Foundation Section. The measurements should be within ¼ inch if the unit is square, if not, loosen the bolts and jockey the sill plates until you get them square.

Once square, tighten the nuts.

WALLS

With the sill plate firmly in place, you are ready to put up the walls. Start with the two side walls. The unit is designed so the plywood lip you left on the inside skin of the walls can be nailed into the sill plate. Position the wall over the sill plate, making sure the plywood lip is firmly against the sill plate. Don't nail any walls in place yet.

Position all four walls on the sill, making sure they fit before nailing any of the bottom lips into the sill plate. Once you have all four walls together, then nail them to the sill plate. Be sure the tops align before nailing.

After nailing the lips to the sill with the barbed nails, use some of the 12d nails and toenail the walls into each other and the sill plate. To do this, remove the outer piece of insulation from the area to be nailed, nail away and then replace the insulation.

Be sure to toenail the corners together tightly, especially the rear wall corners.

FLASHING

The flashing is put on in four separate pieces. Put on both sides first, as they are smaller and somewhat easier to work with. Hopefully you will gain experience on the sides before trying the harder front and back pieces. To put on a side, cut a piece of flashing 56 inches long. This will make it just slightly smaller than the total length of the sill you will be covering.

On the front sill plate, make a series of marks, 1 inch from the edge. These will be used to line up the flashing for bending. Put one edge of the flashing on the marks, and with your helper steadying the flashing, slowly move along the length of the flashing, bending it over the edge of the sill plate. Don't try for a 90 degree bend all at once. The first time just start shaping it with your fingers to get a gentle bend to it. Once the bend is started along the entire length, use two scrap pieces of lumber to bend it as shown. After a few passes the lumber will give you a fairly smooth bend.

With the 1 inch lip bent the entire length of the piece, put the piece in place on the sill, with the lip extending up the side of the bottom plate of the wall. Center the flashing on the sill plate. Nail it to the wall in two or three places, just enough to be firmly in place.

Again, using your hand first, begin to bend the flashing down and over the sill plate, so it covers the insulation.

Using a utility knife, cut the 1 inch lip that is still standing, right at the edge of the wall, so it can be bent flat, and you're done with that piece.

The front and back pieces are somewhat harder, because you have to flash over the corners as well. Cut these pieces so the flashing is about 8 inches longer than the entire length of the sill plate to be flashed. From the outside edge of one sill to the outside edge of the other sill should measure 105 inches. Thus you want your flashing cut 113 inches, or a bit more.

Make the 1 inch bend the same as you did for the sides, and nail the piece in place the same, being sure it is centered. Bend the front edge over the sill and insulation.

To do the corners, start on top of the sill first. Cut the 1 inch lip with a utility knife as shown in the photo, so it lays flat on the sill. Then cut along the fold over the sill so you can bend the front piece back along the side, and the top piece down over the front, as shown in the illustration.

Nail the flashing corners on the edge of the sill, not on the face. Don't use any more nails than absolutely needed. Each nail is a possible place for a leak. The main purpose of the flashing is to keep water from sitting between the bottom plates of the walls and the sill plate and to shield the insulation from sunlight and physical damage.

With the flashing in place on all four sides, nail the outer skins onto the walls.

With this done, you can back fill around the frame, placing dirt up to the top of the sill plate.

With the unit backfilled, but the inside soil cavity not filled, paint all the walls. The inside must be painted white, but the outside can be painted any color you want.

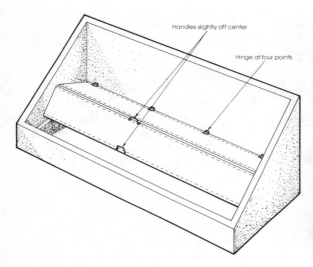

Handles slightly off center

Hinge at four points

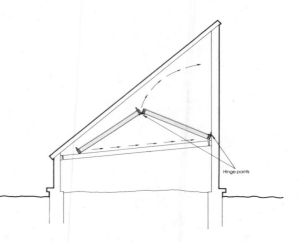

Hinge points

SHUTTER

One of the key design features of the solar growing frame is the night shutter. More than 60 percent of the frame's heat loss would occur through the glazing at night, if there were no shutter. This simple device can add up to six degrees a day to the frame's average temperature.

The shutter must be installed before the top goes on, and after the inside of the frame has been painted. In operation, the shutter folds in two and rests against the back wall during the day. At night, it is unfolded and rests on the shutter support rails.

To make the shutter, start with a 4 x 8 foot sheet of 1 inch thick Thermax sheathing. Cut the sheathing to length with a utility knife and a straight edge. The shutter should measure ½ inch shorter than the inside length of the frame. Measure your assembled frame to find the exact measurement you need. We recommend you make the shutter a full ½ inch shorter than the inside measurement.

To find the needed width of the shutter, measure from the back of the frame to the front of the frame, along one of the side shutter support strips. Subtract ¼ inch from the figure and divide the new measurement by two. That will give you the finished size of each half of the shutter.

If you cut your shutter strips from the piece of 1 x 6 as recommended, subtract 1½ inches from the width measurement for the size of each piece of insulation needed for the shutter. Cut the insulation to the width you have calculated.

Next, cut the four shutter strips to the same length you cut the insulation board. Each of the four shutter strips will go along one of the long sides of the shutter, giving a good surface to fasten hinges to, and giving the shutter added strength and support. Using the aluminum ductape, go around each piece of insulation once, taping the wood to the board. Then turn the board over, and tape from the other side as shown in the illustration. The finished job should have the ductape solidly holding the wooden shutter strips against the sides of the insulation board.

With the wooden strips in place, attach four hinges to one of the strips. Then, attach this strip to the back wall of the grow frame. The insulation board should be resting on the shutter support strips when the hinges are screwed into the frame. With the rear half attached to the wall, raise that section flat against the rear wall and attach four more hinges to the other edge of the insulation board. Then attach the other side of the hinges to the other piece of insulation board, making a hinged unit

Insulation board

Wood shutter strip

Aluminum tape

of the two pieces as shown in the illustration.

Next, fasten a handle to each wooden strip on the front piece of insulation board. Stagger the handles to one side of the center. Lastly, put a strip of regular ductape on the face of the two edges of the insulation board that are hinged together. This will act as an additional seal to keep in heat when the shutter is pulled into place.

With the shutter built and the inside of the frame painted with two coats of paint, you can fill the soil box and get ready to attach the top.

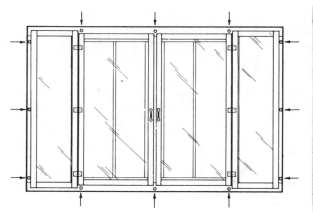

TOP

With the soil bed full, the inside of the frame painted, and the shutter installed, you are ready for the last major step in putting your growing frame together, putting on the top. This is a two person job, no doubt about it. Do not try to do it alone, not only might you hurt yourself, you might damage your growing frame.

The top is put on in two steps. First put the top on the unit, get it exactly in position, and drill pilot holes into the frame. Then you remove the top. With the top off, you put the fastening bolts in the frame, and weather strip the frame. When that is done you work on the top, putting the inner glazing on the frame and doors. If you were to put this thin layer of glazing on while trying to position the top, chances are good one or more of the panels of glazing would tear. Thus, for safety sake, the top goes on twice.

To position the top, have a helper give you a hand lifting it in place. The top should have a slight overhang on the sides, and only about ½ inch of overhang at the top. Once you have it positioned, nail it in place with two nails, you will be able to remove later. Drill holes at the 12 places shown on the diagram. Carefully check to see that each hole will be going into solid wood. Keep the holes as close to the batten strips as you can for the best

results. The hole you drill should be the proper size for the wood screw part of the hanger bolt, ¼ is the right size for a ⅜ X 4 inch hanger bolt.

After drilling the 12 pilot holes, remove the top from the frame. The holes in the frame may have to be drilled a little deeper with the ¼ inch bit. The holes in the top must be enlarged to ⅜ inch. Put a cap nut on a hanger bolt and tighten it into the hole. Don't tighten them all the way down at this time, just enough to hold firm. You don't have to put all the hanger bolts in at this time, just enough to enable you to line the top up on the frame.

Turning your attention to the top, you have to put on the inside layer of glazing. First, glaze the two panels on the top, then the two door units. To put on the glazing, you need to turn the top over, be careful it is well supported so you don't break the latches. Work with care when putting on the glazing, especially if you are hammering, the top is in a pretty vulnerable position at this stage.

Staple the polyethylene to lattice strip E-5, and then roll the plastic around the strip a couple of times, staple the strip on to the frame, being sure to run a bead of caulk on the frame first, where the lattice will be stapled. This will give you a double seal. With the one edge stapled to the frame, pull the plastic tight, run a bead of caulk on the frame, and staple into the caulk. Do the same for the other

frame opening.

The two doors are glazed basically the same way, except each door uses four lattice strips, pieces F-6 and F-7, to secure the plastic. Staple the plastic to all four strips, pull them tight and staple them in place, again putting down a bead of caulk first. This operation goes much faster and with far better results, if you have a helper. To get a good seal, ⁹/₁₆ inch staples work best for putting the material to the frame, but smaller staples are best for fastening the polyethylene to the lattice strips.

With all the glazing on the frame, put a row of foam weather stripping to the outside of the hanger bolts, all the way around the frame opening. A layer of weather strip outside the bolts, and another layer inside the bolts is ideal, but not mandatory. With the weather strip in place, lower the top onto the hanger bolts in the frame. Once the top is on the frame, fasten the remaining hanger bolts tight, and the frame is put together. All that remains is to caulk the unit.

The first area you should caulk is the outside of the glazing surface. A bead of caulk should go all around each of the four glazing areas. Run the bead, and then smooth it out. Use a latex caulk, and wet your finger to smooth the caulking to a tight fit.

Next, go inside the frame and caulk all four corners, as well as all four sides where the top meets

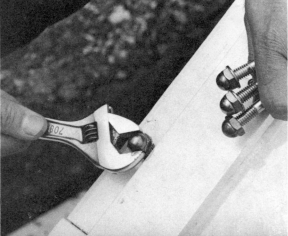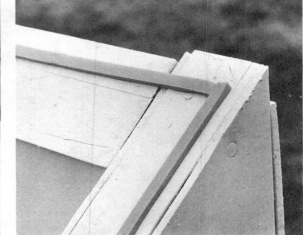

the frame. Any place where there is a large opening, put in extra caulk. Be sure you have caulked all eight seams inside the frame, four corner seams and four seams where the top meets the wall.

Go outside the frame and caulk all the way around where the top sets on the frame. Pay special attention to the back and front walls, where the top meets them. When this is done, caulk the four corner seams and your solar growing frame is finished.